Developments in Electric Discharge Machining

This book covers both fundamentals and recent advancements in Electrical Discharge Machining (EDM) technology, including developments in machine tools, process capabilities, subsystems, electrodes and dielectrics, control units, and more. It also explores EDM variants such as wire-EDM, electrochemical discharge machining, and other hybrid processes. Key topics include miniature parts fabrication, creating complex shapes and features, machining thin components, microfabrication, and using EDM for coating applications.

- Includes both fundamental and advanced topics in EDM
- Discusses micro-EDM, modeling, green manufacturing, and processing of specific materials
- Covers novel approaches such as the hybridization of EDM with other conventional or advanced processes
- Explores case studies on research investigations to provide real-world insights into EDM-based product manufacturing
- Examines monitoring and control, intelligent processing, precision and micromachining, and machining of ceramics and composites

This book is intended for researchers and graduate students in manufacturing technology, mechanical engineering, materials science, and metallurgy.

Advanced Materials Processing and Manufacturing

Series Editor:
Kapil Gupta

The CRC Press Series in Advanced Materials Processing and Manufacturing covers the complete spectrum of materials and manufacturing technology, including fundamental principles, theoretical background, and advancements. Considering the accelerated importance of advances for producing quality products for a wide range of applications, the titles in this series reflect the state-of-the-art in understanding and engineering the materials processing and manufacturing operations. Technological advancements for enhancement of product quality, process productivity, and sustainability, are on special focus including processing for all materials and novel processes. This series aims to foster knowledge enrichment on conventional and modern machining processes. Micro-manufacturing technologies such as micro-machining, micro-forming, and micro-joining, and Hybrid manufacturing, additive manufacturing, near net shape manufacturing, and ultra-precision finishing techniques are also covered.

Additive Manufacturing for Biomedical Implants and Prosthetics
Edited by Pawan Kumar Rakesh and Pravin Kumar

Manufacturing in the Digital Age
Edited by Mohan Kumar Pradhan, Mahendra U. Gaikwad, and Ragavanantham Shanmugam

Fused Deposition Modelling
Technology, Applications, and Developments
Edited by Anshuman Kumar Sahu and Siba Sankar Mahapatra

Manufacturing Tribology
Edited by Jibin T Philip, Chacko Preno Koshy, and MD Mathew

Machine Learning for Advanced Manufacturing
Edited by Nishant Ranjan, Rashi Tyagi, Ranvijay Kumar, Ashutosh Tripathi, and Amit Verma

Engineered Ceramics and Composite Materials
From Fundamentals to Applications
Edited by Neethu Bhaskar, Sathish Kumar Palaniappan, Sanjay Mavinkere Rangappa, and Suchart Siengchin

Additive Manufacturing of Thin-Walled Structures
Edited by Sagar Nikam, Mayur Sawant and Shaun McFadden

Developments in Electric Discharge Machining
Edited by Ranjan Kumar Ghadai, Prosun Mandal and Sujeet Kumar Chaubey

For more information about this series, please visit: www.routledge.com/Advanced-Materials-Processing-and-Manufacturing/book-series/CRCAMPM

Developments in Electric Discharge Machining

Edited by
Ranjan Kumar Ghadai,
Prosun Mandal and Sujeet Kumar Chaubey

CRC Press
Taylor & Francis Group
Boca Raton London New York

CRC Press is an imprint of the
Taylor & Francis Group, an **informa** business

Designed cover image: Ranjan Kumar Ghadai, Prosun Mandal and Sujeet Kumar Chaubey

First edition published 2026
by CRC Press
2385 NW Executive Center Drive, Suite 320, Boca Raton FL 33431

and by CRC Press
4 Park Square, Milton Park, Abingdon, Oxon, OX14 4RN

CRC Press is an imprint of Taylor & Francis Group, LLC

ISBN: 9781032812489 (hbk)
ISBN: 9781041294672 (pbk)
ISBN: 9781003769675 (ebk)

DOI: 10.1201/9781003769675

Typeset in Times
by codeMantra

Contents

Authors' Biography

Ranjan Kumar Ghadai is a distinguished academician with extensive research experience in the field of Materials Science and Nanotechnology. He earned his PhD in Mechanical Engineering and an ME in Production Engineering from the Indian Institute of Engineering Science and Technology, Shibpur, India. Additionally, he holds a B.Tech in Mechanical Engineering from Biju Patnaik University of Technology, Odisha, India. He has over ten years of teaching and research experience. Currently, he is working as Associate Professor in the Mechanical and Industrial Engineering Department of Manipal Institute of Technology, Manipal Academy of Higher Education, Manipal, India. Before joining MIT Manipal, he worked as Assistant Professor in the Department of Mechanical Engineering at Sikkim Manipal University, Sikkim, India. He has published more than 90 SCI/Scopus-indexed research articles. His broad research area includes thin film coating deposition and characterization, process optimization, and multi-criteria decision making. Currently, he is handling two sponsored projects as PI and one as Co-PI. His current h-index is 25, and his i10 index is 41. He has guided three PG students and is currently guiding six Ph.D. scholars. He serves as an Editorial Board Member for Scientific Reports and as an Associate Editor for *Frontiers in Mechanical Engineering.* He has edited eight books and four conference proceedings. Additionally, he is a reviewer for several peer-reviewed journals, including *Journal of Alloys and Compounds, Tribology International, Robotics and Computer-Integrated Manufacturing, Advances in Manufacturing, Surface and Interface*, and *Materials Chemistry and Physics*, among others.

Prosun Mandal obtained his BE in Mechanical Engineering from Jadavpur University in 2010. After that, he joined Cognizant Technology Solutions (MNC). There, he worked as a programmer analyst for two years. He obtained his ME in Production Engineering from the Indian Institute of Engineering Science and Technology, Shibpur, in 2014. After ME, he worked as Assistant Professor in the Department of Mechanical Engineering at Neotia Institute of Technology Management and Science, Diamond Harbour, for nine months. He received his PhD from the Indian Institute of Engineering Science and Technology, Shibpur, in 2020. Then he served as temporary faculty in the Mechanical Engineering Department at Motilal Nehru National Institute of Technology, Allahabad, for nine months. Currently, he is working as Assistant Professor in the Department of Mechanical Engineering of the National Institute of Technology (NIT), Silchar. His research areas include advanced machining, EDM, and composite materials. He has published more than ten papers in highly reputed international journals. He is a reviewer of some international journals like *IJAMT, Journal of Engineering Manufacture, Journal of Process Mechanical Engineering, World Journal of Engineering*, etc.

Sujeet Kumar Chaubey earned his PhD in Mechanical Engineering from the Indian Institute of Technology Indore (IIT Indore). He completed his M.Tech in Manufacturing Systems Engineering from SLIET Punjab (Deemed University) and his B.Tech in Mechanical Engineering from Gautam Buddha Technical University, Lucknow. Additionally, he holds a Diploma in Mechanical Engineering with a specialization in Power Plant from the Institute of Engineering and Rural Technology, Prayagraj. After completing his post-graduation, Dr. Chaubey served as Assistant Professor at Mewar University, Rajasthan, and ABES Institute of Technology, Ghaziabad. Following his PhD, he joined Gautam Buddha University (GBU), Greater Noida, as Assistant Professor (Contract). He then moved to the University of Johannesburg, South Africa, where he worked as a Post-doctoral Research Fellow. Prior to his tenure at GBU, he was awarded the Chinese Postdoctoral Fellowship to carry out post-doctoral research at the Harbin Institute of Technology, Shenzhen, China. Currently, he is serving as Assistant Research Professor at Marwadi University, Rajkot. He has a strong foundation in experimental research and has supervised six M.Tech dissertations and two B.Tech projects. His area of expertise includes advanced manufacturing processes, particularly electrical discharge machining (EDM), wire-electrical discharge machining (WEDM), and hybrid machining techniques. He has authored several research papers published in reputed international journals and serves as a reviewer for numerous journals from prominent publishers, including Springer, Taylor & Francis, Sage Publications, MDPI, Hindawi, and IOPscience.

Notes on Contributors

Shahid Ali, Department of Mechanical & Aerospace Engineering, Nazarbayev University, Astana, Kazakhstan-010000

Cem Alparslan, Faculty of Engineering and Architecture, Department of Mechanical Engineering, Recep Tayyip Erdoğan University, Rize, Turkey-53100

Sameer Kumar Anand, Department of Mechanical Engineering, Indian Institute of Technology Palakkad, Kerala, India-678557

Şenol Bayraktar, Faculty of Engineering and Architecture, Department of Mechanical Engineering, Recep Tayyip Erdoğan University, Rize, Turkey-53100

Dupadu Chakradhar, Department of Mechanical Engineering, Indian Institute of Technology Palakkad, Kerala, India-678557

Rishi Dwivedi, Development Management Institute, Patna, Bihar, India-800001

Amitava Mandal, Department of Mechanical Engineering, Indian Institute of Technology (ISM) Dhanbad, Dhanbad, Jharkhand, India-826004

Prosun Mandal, Department of Mechanical Engineering, National Institute of Technology Silchar, Assam, India-788010

Akash Nag, Faculty of Mechanical Engineering, VSB-Technical University of Ostrava, Ostrava, Czechia-70800

Asma Perveen, Department of Mechanical & Aerospace Engineering, Nazarbayev University, Astana, Kazakhstan-010000

Shanmuga Priyan V.G., Department of Mechanical Engineering, Indian Institute of Technology Palakkad, Kerala, India-678557

Mrityunjoy Roy, Programme of Financial Management, Xavier Institute of Social Service, Ranchi, Jharkhand, India-834001

Abhijit Saha, Department of Mechanical Engineering, Haldia Institute of Technology, Haldia, West Bengal, India-721657

Deepti Ranjan Sahu, Department of Mechanical Engineering, Indian Institute of Technology (ISM) Dhanbad, Dhanbad, Jharkhand, India-826004

Manchu Mohan Krishna Sai, Department of Mechanical Engineering, Indian Institute of Technology (ISM) Dhanbad, Dhanbad, Jharkhand, India-826004

Dipika Sarkar, Department of Computer Science & Engineering (Artificial Intelligence & Machine Learning), Haldia Institute of Technology, Haldia, West Bengal, India-721657

Amresh Kumar Sinha, Sikkim Manipal Institute of Technology, Rangpo, Sikkim, India-737136

Smita, Government Polytechnic, Gulzarbagh, Patna, Bihar, India-800007

Kanmani Subbu S., Department of Mechanical Engineering, Indian Institute of Technology Palakkad, Kerala, India-678557

Preface

Electrical Discharge Machining (EDM) has emerged as one of the most prominent non-conventional machining processes for shaping complex and hard-to-machine materials with high precision. In the past few years, EDM has evolved significantly—from conventional spark erosion processes to intelligent, eco-friendly, and digitally optimized systems. Considering these advancements, this book titled Developments in Electric Discharge Machining aims to bring together recent developments, research innovations, and practical applications that are shaping the future of EDM.

This edited book presents eight carefully selected chapters contributed by prominent researchers across the globe. This book has a diverse range of topics, which include green and sustainable EDM practices, powder-mixed EDM techniques, the use of bio-dielectric fluids, machine learning-based EDM process modeling, optimization strategies using hybrid algorithms, and advancements in tool materials and decision-making models for accurate and precise machining. Special care has been taken to consider the current trends such as sustainability, artificial intelligence, and advanced composite materials, making this book highly useful for the modern manufacturing landscape.

The main motivation behind editing this book is to offer a unified platform that can reflect the multidisciplinary nature of EDM research and its integration with emerging technologies. We firmly believe this book will serve as a valuable resource for graduate and postgraduate students, researchers, academicians, and industry professionals involved in the field of mechanical engineering, manufacturing science, and production technology.

We extend our thankfulness to all the contributing authors for their insightful chapters and to CRC Press for their constant support and encouragement throughout the editorial process. We also appreciate the peer reviewers whose feedback helped enhance the quality and clarity of the content.

We hope that this book will inspire future research and innovation in the field of EDM and contribute meaningfully to the advancement of manufacturing science.

Dr. Ranjan Kumar Ghadai
Associate Professor
Manipal Institute of Technology
MAHE, Manipal, India

Dr. Prosun Mandal
Assistant Professor
National Institute of Technology, Silchar, India

Dr. Sujeet Chaubey
Postdoctoral Fellow, *University of Johannesburg, South Africa*

Introduction

Electrical Discharge Machining (EDM) is a vital non-traditional machining process widely used in modern manufacturing, particularly for shaping hard-to-machine materials, intricate geometries, and thin, delicate components that are often challenging or impossible to produce using conventional methods. Working on the principle of thermoelectric erosion, EDM has evolved into a versatile family of processes—including wire-EDM and powder-mixed EDM (PMEDM)—tailored to meet diverse material and design requirements.

This book, *Developments in Electric Discharge Machining*, presents a comprehensive overview of both the fundamentals and recent advancements in EDM technology. It covers significant developments in machine tools, electrodes and dielectrics, and process optimization techniques.

A key focus of contemporary manufacturing is environmental sustainability. EDM, traditionally reliant on synthetic dielectrics and energy-intensive processes, is undergoing transformation to become more eco-friendly. One chapter addresses Green EDM, highlighting sustainable practices and strategies to reduce environmental impact. Another chapter explores the use of bio-dielectric fluids, offering promising alternatives to conventional hydrocarbon-based dielectrics for cleaner operations.

The addition of powders to the dielectric medium has led to a new variant—PMEDM—which significantly enhances machining performance. A dedicated chapter reviews the evolution of PMEDM and suggests future directions for research and application.

With the advent of Industry 4.0, machine learning (ML) has emerged as a powerful tool for modeling and predicting EDM performance. This book includes a chapter on ML-based predictive modeling, illustrating how data-driven approaches can lead to improved process efficiency and control.

The machining of advanced ceramics and composite materials using EDM is another vital area explored in this volume. These materials, though difficult to machine conventionally, can be effectively processed using EDM with appropriate techniques and tool materials.

Optimization of EDM processes is extensively discussed, including a multi-objective optimization approach using artificial neural networks (ANNs) and non-dominated sorting genetic algorithm-II in wire-EDM, and another chapter focusing on the Taguchi-based Grey Relational Analysis (GRA) technique for optimizing EDM performance using novel Cu-multiwalled carbon nanotube composite tools.

Finally, this book concludes with an exploration of intelligent decision-making in machining using a hybrid criteria importance assessment-proximity indexed value model, applied to precision machining of titanium—a material of high relevance in aerospace, biomedical, and defense industries.

By integrating sustainability, artificial intelligence, novel materials, and process optimization, this book offers a rich source of knowledge for researchers, practitioners, and graduate students working in the field of advanced manufacturing and EDM.

1 Green Electrical Discharge Machining
Sustainable Practices and Environmental Impact Reduction

Cem Alparslan and Şenol Bayraktar

1.1 INTRODUCTION

The environmental and economic challenges brought by the 21st century necessitate the redesign of manufacturing technologies within the framework of sustainability principles. In this context, the adoption of eco-friendly practices in industrial processes is of great importance for both enhancing energy efficiency and minimizing environmental impacts [1,2]. Electrical Discharge Machining (EDM) is a widely used method, particularly for the precise shaping of hard-to-machine materials. However, traditional EDM processes have environmental disadvantages such as high energy consumption and the use of toxic dielectric fluids [3]. These disadvantages constitute a significant obstacle to sustainable manufacturing goals. At this point, the eco-friendly EDM technology called "Green EDM" offers innovative solutions to these problems. Green EDM aims to reduce the negative environmental impacts of traditional EDM through various strategies such as optimizing energy efficiency, using biodegradable and non-toxic dielectric fluids, improving waste management, and increasing overall process efficiency [4]. The adoption of Green EDM not only contributes to environmental sustainability but also increases the economic and technical efficiency of production processes, thereby enhancing industrial competitiveness (Figure 1.1).

Figure 1.1 represents the environmental, social, and economic impact of Green EDM technology. There is a Venn scheme. It includes Green EDM and the subsets it covers.

This chapter provides a comprehensive review of the fundamental principles, applications, and industrial potential of Green EDM. Also, it discusses the future role of eco-friendly manufacturing technologies and the necessary steps for expanding their industrial applications.

DOI: 10.1201/9781003769675-1

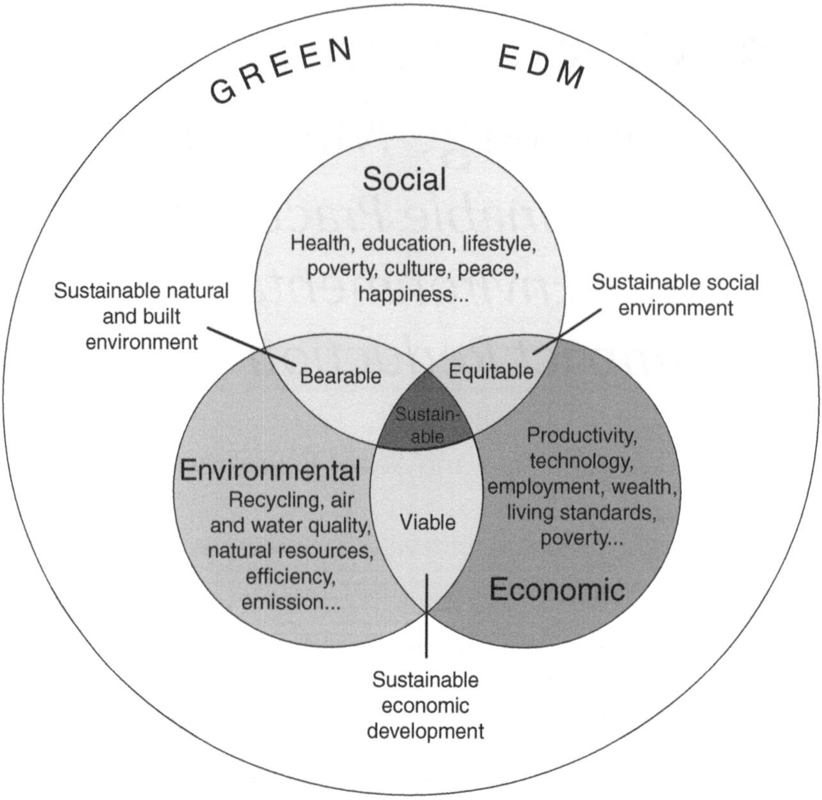

FIGURE 1.1 Environmental, social, and economic impact of EDM.

1.1.1 What Is Green EDM?

Green EDM is a method developed to reduce the environmental effects of traditional EDM technology. While EDM shows superior performance in machining complex and precise workpieces, it also has several negative effects, such as high energy consumption and the use of environmentally harmful machining fluids [5].

In this context, Green EDM aims to minimize the negative environmental impacts of EDM technology by prioritizing sustainability and environmentally friendly applications in manufacturing processes. Its basic components include energy efficiency, eco-friendly dielectric fluids, waste management, and process optimization. It aims to achieve lower energy consumption and higher processing performance with energy efficiency. In this way, both energy costs and the carbon footprint are reduced. The use of eco-friendly machining fluids replaces traditional EDM fluids and offers biodegradable and non-toxic alternatives. These fluids prevent environmental pollution and protect the health of operators. Waste management is another important component of Green EDM. Metal particles and other wastes generated during processing are collected and recycled in a way that will not harm the environment. In this way, material efficiency is increased and environmental impacts are

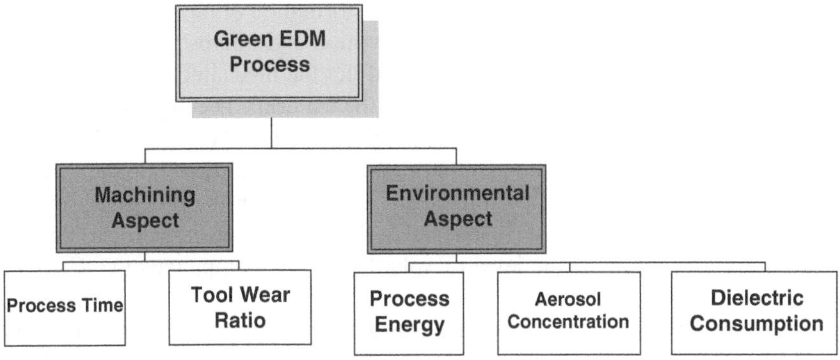

FIGURE 1.2 Environmental and machining aspects of a Green EDM process.

minimized. Process optimization aims to provide the highest efficiency with minimum material loss and surface roughness (SR) by carefully adjusting the processing parameters. The optimization process increases production quality and provides energy and material savings [6].

Figure 1.2 represents the environmental and machining processes of Green EDM. There is a class diagram. It includes machining and environmental processes.

As a result, Green EDM offers an innovative and eco-friendly approach to industrial manufacturing processes, ensuring environmental sustainability and economic efficiency together. This technology both contributes to increasing environmental responsibility awareness and constitutes an important step for the future sustainability of manufacturing processes. Figure 1.2 shows the processing and environmental aspects of Green EDM. The dissemination of Green EDM plays a critical role in the adoption of environmentally friendly practices in industrial manufacturing.

1.1.2 Environmental Effects and Unsustainability of Traditional EDM

EDM technology is especially preferred for machining difficult-to-machine materials such as hardened steels, superalloys, and composite materials. However, these technological advantages provided by EDM create significant environmental and sustainability problems throughout the process. One of the most obvious environmental impacts of traditional EDM is high energy consumption. Electrical energy is constantly used to create spark discharges between two electrodes throughout the process. Since this energy consumption is mostly provided by electricity obtained from fossil fuels, it leads to significant carbon emissions and energy waste. This process, which has low energy efficiency, does not comply with global sustainability goals. Another important problem is the waste generated during the EDM machining [7]. Micro particles broken off from the parts and electrode mix with the machining fluid and create waste. Since the machining fluids used in the EDM process often contain toxic and environmentally harmful chemicals, proper management of the resulting waste is critical for environmental protection [8]. Therefore, the lack of appropriate waste management systems causes environmental pollution in the traditional EDM method. In addition, the formation of high temperatures during the EDM process leads to thermal deterioration and microstructural changes in the workpiece. This situation can negatively affect the mechanical properties of the processed material and reduce the usability of the material. The high consumption of electrode material also creates a negative situation in terms of resource efficiency and threatens sustainability goals. Finally, harmful emissions released into the air during EDM machining negatively affect the health of workers in the processing environment and harm the environment. In particular, the evaporation of hydrocarbon-based machining fluids increases the level of volatile organic compounds (VOCs) in the atmosphere, causing air pollution.

These environmental impacts and unsustainable practices of traditional EDM reveal the need to develop greener and more eco-friendly manufacturing technologies. In this context, Green EDM technologies have important potential to minimize environmental impacts and provide sustainable manufacturing processes [4].

1.1.3 The Importance and Advantages of Green EDM

The environmental effects and unsustainability of traditional EDM have necessitated the adoption of more eco-friendly approaches in manufacturing technologies. Green EDM technologies respond to these needs by providing energy efficiency, waste management, reduction of thermal degradation, and healthier working environments [5]. The method has many advantages, some of which are detailed below.

Energy efficiency: Green EDM technologies include innovations aimed at increasing energy efficiency. Unlike traditional EDM, it provides the same machining precision and quality with lower energy requirements. This helps notably reduce the carbon footprint. Energy efficiency, which reduces dependence on fossil fuels, both reduces costs and contributes to environmental sustainability [9,10].

Environmentally friendly dielectric fluids: In these methods, biodegradable and eco-friendly alternatives are used instead of the toxic and environmentally harmful dielectric fluids traditionally used in EDM. These fluids minimize air pollution and reduce environmental impacts by preventing the spread of VOCs [11].

Waste management and recycling: It focuses on minimizing and recycling waste generated during the machining process. Waste generated as a result of micro particles and molten material mixing with the processing fluid can be recycled and reused when managed correctly. Thus, resource efficiency is increased, and environmental issues related to waste disposal are reduced.

Reduction of thermal degradation: It minimizes thermal degradation and microstructural changes in the workpiece. This helps maintain the mechanical properties of the machined material, thereby increasing material efficiency. Thermal deterioration, which is frequently encountered in traditional EDM, reduces the quality of the material. Green EDM minimizes such negative effects by using strategies such as the use of conductive fluid, low energy pulses, enhanced heat distribution, microwave-assisted machining, pulse duration optimization, and the correct selection of electrodes and materials.

Improvement of the working environment: It provides a safer and healthier working environment by reducing harmful emissions and toxic chemicals in the machining environment. This protects worker health and increases the overall safety of the work environment. In particular, reducing VOCs and other harmful gases minimizes long-term health risks for workers.

Sustainable manufacturing practices: It facilitates the integration of sustainable manufacturing practices. Thus, while reducing the environmental impact of manufacturing processes, it helps businesses achieve their eco-friendly and sustainability goals. Green manufacturing technologies provide a competitive advantage, making them attractive to environmentally conscious consumers and investors [5,12].

1.2 PRINCIPLES OF GREEN EDM

Green EDM technologies are based on several principles, such as energy efficiency, waste reduction and recycling, and optimizing water and oil usage [13]. These principles are critical to minimizing environmental effects in machining processes and ensuring more sustainable industrial production. Green EDM not only provides environmental benefits but also offers economic advantages to businesses. It increases competitiveness by reducing machining costs with strategies such as energy efficiency and waste reduction. In addition, companies that adopt environmentally friendly production methods in today's industry achieve sustainability goals and become more attractive to environmentally conscious consumers and investors.

Green EDM plays a key role in the adoption of eco-friendly and sustainable manufacturing technologies. These principles contribute to minimizing environmental effects in industrial manufacturing and building a more sustainable future.

1.2.1 Working Principles

The working principle of Green EDM technology is based on similar principles to traditional EDM. However, it is optimized in terms of environmental and energy efficiency. Green EDM performs the material machining by creating high-frequency electrical discharges between two electrodes. This process is based on the formation of small sparks between the workpiece and the electrode and the removal of material from the workpiece by means of these sparks (Figure 1.3) [14,15]. The main difference in the working principle of Green EDM is the integration of technological innovations that minimize energy consumption and environmental effects. This allows for more efficient use of energy resources and reduction of waste generated during the machining. High-efficiency energy sources and innovative power sources are used in Green EDM to increase energy efficiency. This reduces the amount of energy required during the manufacturing while maintaining the efficiency of the machining process [16,17]. Another important element is that the machining fluids used are environmentally friendly and biodegradable. Instead of the toxic machining fluids used in traditional EDM, Green EDM minimizes environmental pollution by using water-based or eco-friendly oil-based fluids. These machining fluids perform cooling and washing functions during the process while reducing environmental damage [18–20].

Figure 1.3 represents the cutting process diagram in EDM. There is one image. It includes different equipment for EDM.

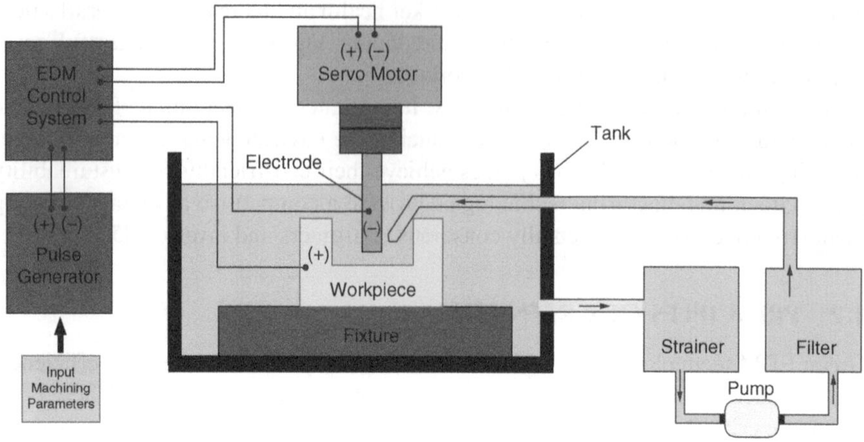

FIGURE 1.3 Schematic diagram of EDM.

In addition, the working principle of Green EDM includes waste management and recycling strategies. The micro particles and molten materials formed during machining are collected and recycled through advanced filtration systems. This significantly reduces the amount of waste and increases material efficiency [5]. Finally, Green EDM technologies minimize thermal distortions by optimizing processing parameters. High-frequency and controlled discharges reduce thermal impacts on the workpiece, preserving the microstructural integrity of the machined material and preventing deterioration of its mechanical properties. These working principles provide a more sustainable and eco-friendly machining process.

1.2.2 Energy Efficiency in EDM

One of the main goals of Green EDM technologies is to increase energy efficiency. While traditional EDM machining is known for its high energy consumption, Green EDM reduces environmental effects and costs by ensuring more efficient use of energy resources. Energy efficiency is achieved through many different methods in Green EDM technologies [5,9]. First of all, high-efficiency energy sources and innovative power sources are used. These power sources optimize energy consumption and minimize the amount of energy required in the manufacturing process. High-frequency and controlled discharges reduce energy loss and increase the efficiency of the manufacturing process. In addition, energy recovery technologies used in Green EDM systems play an important role in increasing energy efficiency. The recovery and reuse of heat generated during manufacturing notably reduce total energy consumption [10]. An example of the energy usage of Green EDM and traditional EDM methods is shown in Figure 1.4.

Figure 1.4 compares the traditional and Green EDM machining methods. There is a bar chart. It shows the energy consumption of the two methods.

Another critical element is optimizing the process. Green EDM optimizes the processing parameters (e.g. discharge time, current density, and frequency) to reduce

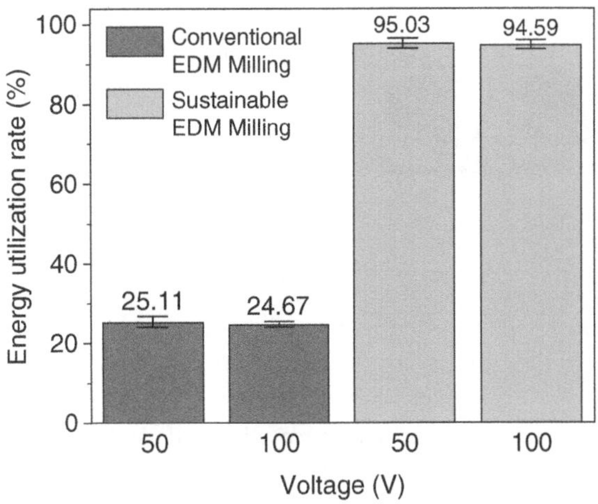

FIGURE 1.4 Comparison of the energy utilization rate between traditional and Green EDM machining methods [21].

energy consumption while maintaining machining speed and quality. This optimization increases energy efficiency while ensuring the efficiency and accuracy of the manufacturing process [5]. Additionally, the eco-friendly machining fluids used in Green EDM also contribute to energy efficiency. These fluids perform cooling and flushing functions with lower energy consumption. Finally, energy monitoring and management systems are also a significant part of Green EDM technologies. These systems continuously monitor and analyze energy consumption and make the necessary adjustments to increase energy efficiency. In this way, energy consumption can be optimized, and unnecessary energy use is prevented. Green EDM makes significant contributions to environmental sustainability with these innovative approaches to energy efficiency. Lower energy consumption ensures both a reduction in environmental impacts and lower machining costs. Thus, it offers a more sustainable and competitive manufacturing process.

1.2.3 Waste Reduction and Recycling Strategies

Green EDM technologies aim to minimize the environmental impacts of machining processes by adopting innovative and eco-friendly approaches to waste management. Waste reduction and recycling strategies constitute the basic components of these processes and play a critical role in achieving sustainable machining goals [4].

Figure 1.5 shows the waste reduction and recycling strategies in the Green EDM method. There is a workflow diagram. It shows the stages through which the waste generated by the parts processed with EDM is recycled.

Waste reduction: Another important component of Green EDM technologies is zero-waste processing techniques. Waste reduction strategies focus on minimizing material waste generated during machining processes. Micro particles and molten materials generated during processing can create significant amounts of waste

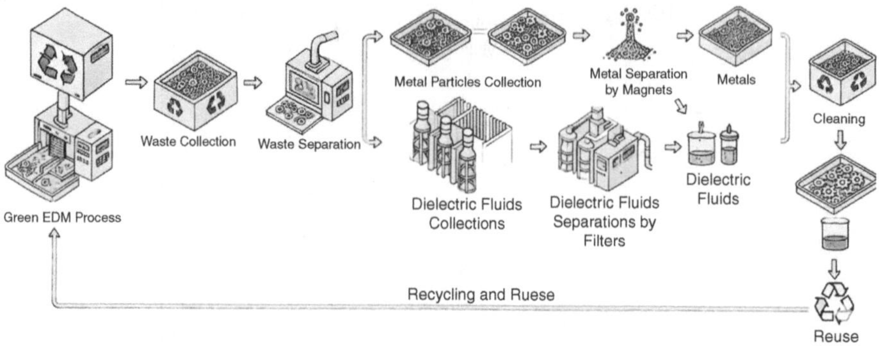

FIGURE 1.5 Waste collection, separation, recycling, and reuse in the Green EDM method.

in traditional EDM. Green EDM uses advanced filtration and separation systems to reduce this waste. These systems effectively filter the waste accumulated in the machining fluid, increasing the reusability of the fluid and minimizing the amount of waste [5]. In addition, since the machining fluids used in Green EDM processes have more eco-friendly and biodegradable properties, the risk of environmental pollution when disposing of the fluids as waste is significantly reduced [22].

Recycling strategies: Green EDM increases resource efficiency by ensuring that waste materials generated during manufacturing are recycled and reused. Metal particles and molten materials generated during machining are processed through advanced collection and recycling systems. These systems collect, separate, and prepare waste materials for reuse. In this way, the efficient use of raw materials is ensured, and environmental problems related to waste disposal are minimized [23]. The machining fluids used in Green EDM processes are also recyclable. Machining fluids are cleaned and prepared for reuse through advanced filtration systems. This importantly reduces fluid consumption and the amount of waste fluid. Green EDM technologies minimize the environmental effect of manufacturing processes and increase resource efficiency with waste reduction and recycling strategies [24]. These approaches both ensure environmental sustainability and offer economic advantages by reducing production costs. Advanced waste management and recycling technologies enable Green EDM to offer an eco-friendly and sustainable machining process.

1.2.4 Water and Oil Usage in EDM

Green EDM technologies increase environmental sustainability and maximize machining efficiency by optimizing water and oil usage. Thus, it minimizes water and oil consumption and reduces the environmental effect of these resources [25,26].

Water usage: Water is widely used as a cooling and machining fluid in EDM processes. Green EDM optimizes water usage, reducing water consumption and minimizing environmental effects [27,28]. The following strategies are implemented to improve water usage:

- The water used during manufacturing is cleaned and prepared for reuse through advanced filtration and purification systems. This importantly reduces water consumption and minimizes the amount of wastewater.
- It uses innovative technologies that reduce water consumption and enable more efficient use of water. These technologies enable more effective use of water as a cooling and machining fluid.
- Monitoring and managing water use play a critical role in optimizing water consumption. Water management systems continuously monitor and analyze water use, making necessary adjustments to minimize water consumption.

Oil usage: The oils used in EDM processes serve cooling and lubrication functions. Green EDM improves oil consumption and environmental effects by using eco-friendly and biodegradable oils. The following strategies are implemented to optimize oil usage:

- Instead of toxic and environmentally harmful oils used in traditional EDM, Green EDM uses biodegradable and eco-friendly oils. These oils decrease environmental pollution and toxic waste.
- Used oils are cleaned and prepared for reuse through advanced filtration systems. Thus, oil consumption and waste oil amount are significantly reduced.
- Green EDM optimizes lubrication systems, minimizing oil consumption. These systems ensure that oil is used more efficiently during processing and prevent unnecessary oil consumption.

Green EDM provides both environmental sustainability and process efficiency by optimizing water and oil usage. These optimization strategies help preserve natural resources while offering economic advantages by reducing manufacturing costs.

1.3 TECHNOLOGICAL INNOVATIONS IN GREEN EDM

Green EDM technologies include significant technological innovations in line with the goals of sustainable manufacturing and minimizing environmental effects. These innovations have been developed to increase energy efficiency, minimize waste production, and use natural resources more effectively. Green EDM's technological innovations include high-efficiency EDM machines and equipment, zero-waste processing techniques, and sustainable material selection and machining methods [22,29]. In this section, the technological innovations of Green EDM and their contributions to sustainable manufacturing are discussed.

1.3.1 High-Efficiency EDM Machines and Equipment

Green EDM technologies aim to reduce energy consumption and increase machining efficiency by using high-efficiency EDM machines and various equipment. These machines and equipment have advanced technological features and support environmental sustainability [30,31]. Some high-productivity EDM machines and equipment are shown in Figure 1.6.

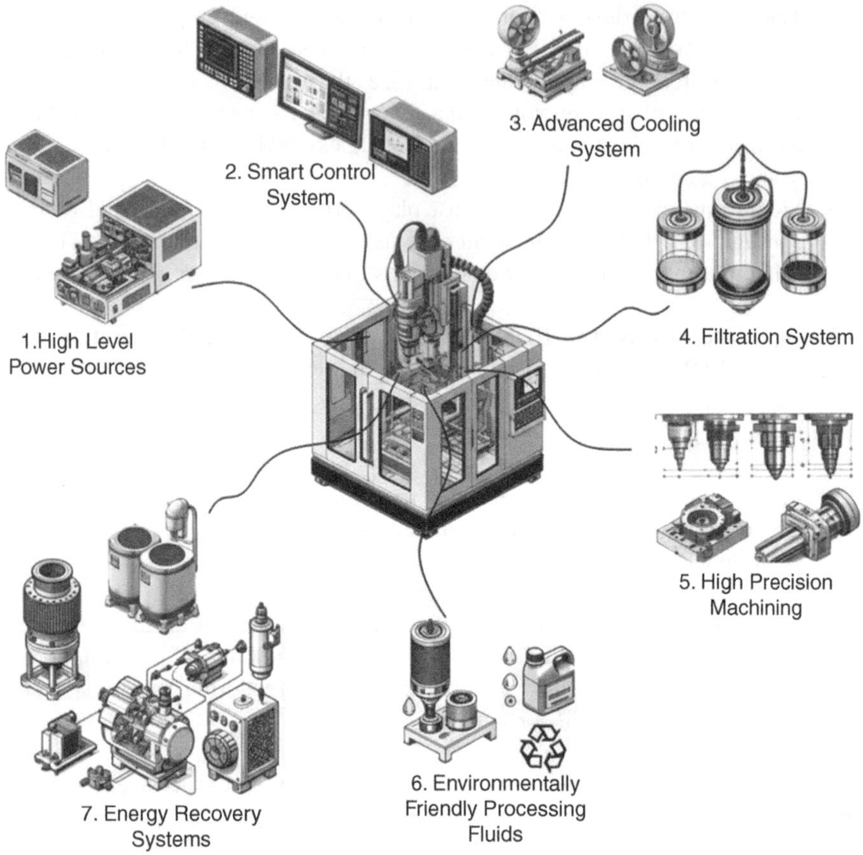

FIGURE 1.6 High-efficiency EDM machines and equipment.

Figure 1.6 represents high-efficiency machines and equipment used in Green EDM. There is an image. In the middle is the Green EDM machine, and around it are seven different types of machine equipment.

Advanced power supplies: These power supplies enhance energy efficiency by minimizing energy losses and enabling more controlled electrical discharge processes. By optimizing process parameters, they reduce energy consumption while maintaining process quality [32]. Adaptive control, nano-pulsed, high-frequency pulse, eco-friendly, and intelligent power supplies are widely used for this method. For example, the use of nano-pulsed power supplies increases energy efficiency by producing short pulses in the nano-second range.

Smart control systems: These continuously monitor and analyze process parameters and automatically make necessary adjustments for optimum performance. These systems increase the efficiency of the process while minimizing energy consumption and provide real-time data to operators [33,34]. In addition, waste monitoring and management systems play a critical role in implementing zero-waste machining techniques. These systems optimize waste management strategies by continuously monitoring and analyzing waste generated during the machining process. Waste monitoring and management systems prevent unnecessary waste generation and ensure that recycling processes are managed effectively. For example, a waste monitoring system equipped with Keyence SJ-H series air ionizers and Baumer LBFS series liquid level sensors continuously monitors the cleanliness of the dielectric fluid and ensures that recycling processes are managed effectively.

Advanced cooling systems: Advanced cooling systems provide efficient removal of heat generated during the machining process. Cooling systems minimize thermal distortions of the workpiece, improving process quality. They also extend the life of cooling fluids, reducing waste generation. For example, the ChillerWorks CW-5000 cooling system is an advanced cooling solution used in Green EDM processes. This system precisely controls the temperature of the dielectric fluid, ensuring rapid heat removal during machining. As a result, thermal damage to the workpiece is minimized, and the lifespan of the cooling fluid is extended, thereby reducing waste.

Advanced filtration systems: These effectively filter waste accumulated in processing fluids and enable the reuse of the fluids. These systems allow for the continuous cleaning and recycling of machining fluids, significantly reducing waste generation. Filtration systems extend the life of the machining fluid and minimize environmental pollution [35]. For instance, the Ebbco Evolution series filtration systems offer an advanced solution. This system effectively filters micro particles and waste accumulated in the dielectric fluid, thereby increasing the reusability of the fluid and reducing the amount of waste.

High-precision machining units: These machining units offer micron-level accuracy. They enable the machining of complex geometries and fine details. These units reduce material waste and increase the quality standards of the machined parts [36,37]. A notable example is the Mitsubishi EA8PV Advance model, a high-precision machining unit. This unit provides micron-level accuracy, allowing for the machining of intricate and detailed parts. It minimizes material waste and maximizes the quality standards of the parts produced.

Environmentally friendly machining fluids and oils: These fluids and oils are biodegradable and non-toxic. Their use prevents the formation of environmentally harmful waste during the machining process while providing effective cooling and lubrication [38–40]. Commercial fluids for EDM include kerosene, hydrocarbon, mineral oil, and deionized water (commonly used with wire EDM). The life of the dielectric fluid depends on various factors such as the type and efficiency of EDM fluid filters, making it difficult to specify a definitive expiration date. In general, if the operator is using an oil-based fluid and it is older than five years, it should be replaced. Since particle suspension is a key feature of machining efficiency, the need to replace the dielectric fluid is determined using a refractometer.

Energy recovery systems: These systems enable the recovery and reuse of heat generated during the machining process. These systems reduce total energy consumption by minimizing energy losses.

These equipment are the basic components of Green EDM technologies and provide significant benefits in terms of energy efficiency, process precision, and environmental sustainability. High-efficient EDM machines and equipment enable modern manufacturing processes to be more efficient and eco-friendly.

1.3.2 SUSTAINABLE MATERIAL SELECTION AND MACHINING METHODS

Sustainable material selection and machining methods play a critical role in ensuring environmental sustainability in Green EDM technologies. These approaches are implemented to minimize the environmental impact of materials used in manufacturing processes and to develop more sustainable production methods. Sustainable material selection and machining methods contribute to the conservation of natural resources, reduction of waste, and improvement of energy efficiency [40,41]. Sustainable material selection is one of the fundamental elements of Green EDM technologies. This approach includes the use of materials that have low environmental impact and are recyclable and biodegradable. In sustainable material selection, features such as recyclability, biodegradability, and low carbon footprint are important.

Recyclable materials: Using recyclable materials in manufacturing processes increases resource efficiency by reducing waste. Recyclable materials can be reused in the production cycle and minimize environmental impacts [5]. For instance, metals such as aluminum and copper are frequently used in EDM processes and have high recycling rates [42].

Biodegradable materials: At the end of their lifecycle, these materials can decompose through natural processes without harming the environment. The use of such materials reduces waste management issues and supports environmental sustainability [43,44]. Polymers and natural composites are among these types of materials.

Low carbon footprint materials: Preferring materials with a low carbon footprint in production processes reduces carbon emissions and contributes to the fight against climate change. These materials are obtained with production methods that are highly energy-efficient and have low environmental impacts. For example, recycled metals and polymers obtained from renewable resources have a low carbon footprint [45,46].

Sustainable machining methods are techniques designed to increase energy efficiency and reduce environmental impacts in production processes [22]. In Green EDM technologies, key factors influencing sustainability include advanced power supplies, advanced cooling and filtration systems, energy-efficient machining techniques, waste management systems, the integration of recycling systems, optimized machining parameters, and the use of eco-friendly dielectric fluids [41,47,48].

1.4 CASE STUDIES

Green EDM technologies offer different solutions to achieve environmental sustainability and energy efficiency goals in various industries. The aerospace industry has adopted Green EDM technologies for the production of components requiring high precision and material durability. For example, in the machining of titanium and nickel alloys, Green EDM has been observed to provide lower energy consumption and waste production compared to traditional EDM methods [49,50]. The automotive sector also utilizes Green EDM technologies to enhance energy efficiency and material recycling in production processes. Specifically, in the machining of aluminum and steel parts, Green EDM methods have been reported to reduce energy consumption and minimize waste generation [9]. These technologies help automotive manufacturers achieve their sustainability goals. In medical device manufacturing, precise processing of biocompatible materials is required. Green EDM enables high-precision processing of materials such as titanium and stainless steel used in this field [51]. The applicability and advantages of Green EDM technologies are supported by various industrial and academic studies. In this section, studies on Green EDM's energy efficiency, waste reduction, sustainable material use, and environmental impact reduction are examined. These studies illustrate how Green EDM is applied in different industries and the results achieved.

Studies examining the effects of different bio-dielectric oils and machining parameters in EDM methods have revealed the significant impact of various oils and parameter settings on machining performance. Das et al. compared the effects of Neem, Canola, and Jatropha-based bio-dielectric oils on material removal rate (MRR), SR, and cutting tool. They observed that Neem oil provided 18% higher overcutting and 40% higher tool wear than other oils. Additionally, Canola oil produced 12% lower SR compared to Neem and Jatropha oils, while Jatropha oil achieved 15% higher MRR [52]. These findings are consistent with the studies conducted by Singh et al. on Pongamia oil. They found that Pongamia oil offers higher MRR, lower electrode wear rate (EWR), and better surface quality compared to Jatropha oil [53]. Mishra and Routara investigated the EDM performance of Polanga bio-dielectric oil compared to standard hydrocarbon dielectrics. They found that this oil increased the average MRR by 0.86 times and improved the surface quality by 16.64%. Also, they found that it increased the surface hardness by 6.46% and emitted 17.33% fewer aerosols [19]. These findings are supported by the studies of Bains et al., who observed that increased spark discharge energy accelerates the melting and evaporation of the workpiece material, thereby providing a higher MRR [54]. Similarly, Yeo et al. reported in their study that lower energy levels produced less molten material, thus reducing SR and burr width [17].

Klocke et al. examined traditional EDM milling for machining titanium (Ti-6Al-4V) and nickel (Inconel 718)-based alloys. They found that the EDM method improved productivity and reduced cost compared to mechanical milling. However, they observed that the machining efficiency was still very low. For instance, they noted that the machining efficiency for Ti-6Al-4V alloy did not exceed 240 mm^3/min, which falls far short of industrial production requirements [55]. Hourmand et al. highlighted in their studies that micro-EDM milling technology has potential in the production of micro-components but faces difficulties in industrial use due to several disadvantages, such as low machining efficiency, high energy consumption, severe electrode wear, and a poor machining environment [56]. These disadvantages were supported by studies by Kou and Han. They showed that high-speed EDM milling with moving direct current electric arcs improved machining efficiency to some extent using water-based dielectrics. However, they still observed that it did not meet the actual machining requirements of industrial production [57].

Several studies examining the optimization of EDM parameters and their effects on materials have been conducted. Wuyi Ming et al. optimized the cutting parameters to minimize energy consumption and exhaust emissions during the EDM machining of Al(6061) and SKD(11). They investigated the effects of three different cutting parameters (peak current, pulse duration, and duty cycle) on MRR, surface roughness (Ra), energy efficiency per volume (EEV), and exhaust emission characteristics (EEC). Initially, they analyzed process energy and exhaust emissions. The results showed a negative correlation between surface quality and EEV for both Al(6061) and SKD(11) materials. Second, they used two data analysis methods (adaptive network-based fuzzy inference system (ANFIS) and Taguchi analysis) to obtain high MRR and best surface quality, EEV and EEC for desired machining performances, and environmental sustainability. They had observed that the relative errors between the predicted values and experimental results for MRR, Ra, EEV, and EEC were no more than 20%, with average relative errors at a maximum of 13.04%. Finally, they had developed a framework implemented by ANFIS, enhanced cuckoo search (ECS) algorithm, and desirability function-based optimization system to achieve the desired cutting parameters. High peak current (8.94 A) and pulse period (500 μs) are proposed for machining Al 6061, and it has been observed that the optimum solutions of EEV and EEC are significantly reduced by 87.13% and 37.09%, respectively, with these parameters. Similarly, high peak current (8.77 A) and pulse period (261.5 μS) are suggested for machining SKD 11, and it is determined that the optimum solutions of EEV and EEC are significantly reduced by 93.73% and 46.85%, respectively, with these parameters. Based on these findings, they concluded that the proposed technique offers significant advantages and potential for applications in the field of green manufacturing [9]. Prabhu and Vinayagam investigated the SR and microcracks on AISI D2 tool steel workpieces using an EDM method that mixed single-walled carbon nanotubes into the dielectric fluid. As a result of the experimental study, it was observed that the samples with carbon nanotubes in the dielectric liquid had reduced microcracks, better surface quality, and improved surface morphology compared to the samples processed by EDM in the dielectric liquid without carbon nanotubes [58]. De Jesus et al. examined the effect of the EDM process on the fatigue behavior of AISI D2 tool steel. The notch impact behavior, residual stress, and fatigue resistance of the material were assessed on smooth and notched specimens and compared with other commercial machining techniques. As a result of the research, they found that, compared to the base material which exhibited compressive residual stress, EDM processing led to the formation of tensile residual stresses that adversely affected fatigue resistance. When comparing EDM to grinding processes, they observed that while grinding induced compressive stress on the material surface, EDM processing caused both the appearance of cracks and the formation of tensile residual stress in the recast layer, also known as the re-solidified layer [59]. Pragadish and Kumar performed EDM and dry EDM machining on AISI D2 tool steel. They used current (8, 10, and 12 A), pulse on time (100, 150, and 200 μs), gap voltage (50, 60, and 70 V), pressure (1.5, 2.05, and 2.5 KPa), speed (1,500 rpm), and duty factor (32) as machining parameters. Copper electrodes were used for both methods, and they examined the effects of these two different machining methods on MRR, SR, and surface morphology. The study concluded that the samples machined with dry EDM showed better MRR and surface quality compared to those machined with traditional EDM, with an improvement of 16.8% in surface quality [60]. Abbas et al. aimed to improve the surface quality of AISI D2 steel machined by EDM using

Span-20, a surfactant, and nano chromium powder (NCP) in the dielectric fluid. As a result of the study, they observed that the average SR for steel improved when the NCP concentration and current pulse duration were at maximum levels and when Span-20 was used [61]. Khan et al., by changing the parameters of current (1, 4, and 7 A), voltage (100, 125, and 150 V), and pulse duration (10, 15, and 20 μs) in EDM machining of D3 steel, tried to determine the ideal values for MRR, tool wear rate (TWR), and SR. To obtain these data, they calculated the signal-to-noise (S/N) ratio using ANOVA and the Taguchi L_9 method. As a result of the study, they achieved ideal values for MRR (current: 7 A, pulse on time: 20 μs, voltage: 125 V), TWR (current: 1 A, pulse on time: 10 μs, voltage: 100 V), and SR (current: 1 A, pulse on time: 10 μs, voltage: 150 V). Regression analysis and R^2 values were calculated as 92.5%, 80.4%, and 89.9% for MRR, TWR, and SR, respectively [62]. Hamid and Lajis, in their studies, used copper tungsten (Cu35%-W65%) produced with powder metallurgy, known for its high wear resistance, as the electrode material in EDM to machine D2 tool steel hardened in gas oil. After the experimental study, they calculated the necessary parameter values (peak current and pulse duration) to achieve the lowest TWR (0.017 mm^3/min) and the highest MRR (43.069 mm^3/min) values [63]. Singh et al. investigated the effects of a cryogenically cooled copper electrode on EDM machining of Inconel 601 material. They used pulse on time (100, 150, and 200 μs), gap voltage (60, 70, and 80 V), and peak current (6, 8, and 10 A) as input parameters. MRR, TWR, and SR values were determined as output parameters. In their experimental study, they used the Box-Behnken design for optimization. The research results indicated that the current parameter significantly affected MRR, TWR, and SR. They observed that the cryogenically cooled electrode improved the surface quality and MRR values of the workpiece. Additionally, they found that, compared to a normal electrode, the cryogenically treated electrode reduced TWR by up to 33% [64]. Prakash et al. investigated the machinability of magnesium AZ31B, a lightweight implant material, using micro-EDM. They used voltage (160 V), pulse on time (30 μs), speed (300 rpm), and capacitance (100 μF) as machining parameters. They also used two different types of tungsten carbide electrodes, cryogenically treated and untreated, during the machining of AZ31B implant material. As a result of the study, they observed that the cryogenically treated tungsten carbide electrode increased the MRR, improved machining time, and resulted in better dimensional accuracy of the workpiece [65]. Anand et al. investigated the machining of high chromium tool steel using the EDM process. They selected peak current (2, 4.5, and 7 A), pulse duration (10, 255, and 500 μs), dielectric level (40, 60, and 80 mm), and flushing pressure (0.3, 0.5, and 0.7 kgf/cm^2) as control parameters. The Taguchi approach was used to optimize the machining parameters. As a result of the study, they found that increasing the peak current led to a reduction in process time and energy consumption. Additionally, they observed that increasing the peak current and pulse duration resulted in an increase in MRR due to the reduction in process time. They noted that output parameters were significantly influenced by peak current and dielectric level, while pulse duration had the least effect [66]. Rao et al. determined the optimal machining parameters for EDM machining of AISI D2 steel using an 8-mm diameter copper electrode with the TOPSIS (Technique for Order of Preference by Similarity to Ideal Solution) method. They optimized the parameters of pulse on time (300, 400, and 500 μs), peak current (6, 7, and 8 A), and servo voltage (40, 50, and 60 V) using the Taguchi L_9 experimental setup. Three levels of each parameter were used to design the experiments. They used non-volatile sunflower oil as the dielectric fluid. MRR, TWR, and SR values were determined for each experiment. ANOVA was used to support the effect of each input parameter on the output responses. They used the Analytical Hierarchy Process (AHP) to select the weight of each response. The TOPSIS method was used for multi-criteria decision analysis to define the ideal parameters for machining AISI D2 steel. According to the AHP-TOPSIS results, they identified pulse on time (500 μs), peak current (7 A), and servo voltage (50 V) as the ideal values [67]. Gopalakannan et al. determined that the percentage effect of peak current, pulse off time, and pulse on time on MRR while machining Al 6063/10% SiC-35% Gr hybrid metal matrix composite is 84.05%, 7.279%, and 5.665%, respectively. This indicates that peak current and pulse off time are significant factors affecting MRR. For TWR, they observed that pulse off time and pulse on time contributed 68.38% and 20.98%, respectively. They also found that peak current, pulse on time, and voltage are important parameters for SR with 43.88%, 38.86%, and 8.35% contribution, respectively [68]. Kumar et al. investigated the effect of conventional and powder metallurgy electrodes on the characterization of machined surfaces using EDM. They also added powder particles to the dielectric medium to enhance process efficiency. They determined the impact of particle presence in the dielectric fluid on surface characteristics and machining improvements (MRR, SR, crater formation, microhardness, etc.) for both conventional and composite materials. Additionally, they observed that particles mixed into the dielectric medium frequently bonded to the molten metal of the workpiece during machining. They found that particles moving from the compacted electrode to the workpiece reduced microcracks (by 40%), voids (by 35%), and the formation of the recast layer (by 50%) on the material surface during the machining process. Thus, they found that the EDM-processed surface had comparatively fewer defects. In EDM processes, they observed that using a powder-mixed electrode reduced SR and increased MRR, while using a composite electrode reduced electrode wear and improved surface quality. When both methods were used together, they achieved the highest surface quality, with minimal recast layer formation and microcracks. Finally, they determined that the microhardness, SR, wear resistance, and corrosion behavior of the material could be improved using these modification techniques [3].

1.5 CONCLUDING REMARKS

Green EDM technologies represent an important innovation for increasing the environmental sustainability of modern production processes. While traditional EDM is characterized by high energy consumption, significant waste production, and environmental impacts, Green EDM technologies are designed to increase energy efficiency, minimize waste, and use more sustainable materials. These principles make manufacturing processes more efficient, economical, and eco-friendly. Technological innovations such as high-efficiency EDM machines and equipment, advanced power supplies, smart control systems, and advanced cooling and filtration systems make the advantages of Green EDM even more evident. The sustainability of Green EDM is supported by methods such as zero-waste machining techniques, recycling and reusing, advanced filtration systems, optimized machining parameters, eco-friendly dielectric fluids, and closed-loop manufacturing systems. Sustainable material selection and machining methods are another important component of Green EDM technologies. The use of recyclable, biodegradable, and low carbon footprint materials minimizes the environmental impact of manufacturing processes, while energy-efficient machining methods ensure the efficient processing of these materials.

The case studies presented demonstrate the applicability and benefits of Green EDM in various industries. In sectors such as aerospace, automotive, medical devices, electronic components, and green energy systems, the use of Green EDM technologies has achieved notable results by reducing energy consumption, minimizing waste, and enhancing environmental sustainability. Recent EDM studies in the literature also show that Green EDM technologies are continuously evolving in the areas of energy efficiency, waste reduction, and sustainable material usage, making significant progress in these fields. Research indicates that Green EDM will continue to gain wider acceptance in industrial applications and further strengthen its potential to enhance environmental sustainability.

In conclusion, Green EDM technologies represent a significant innovation in making modern manufacturing processes more sustainable, eco-friendly, and cost-effective. Energy efficiency, waste reduction, sustainable material use, and technological innovations form the basis of the advantages provided by Green EDM. By enhancing environmental sustainability in industrial manufacturing processes, these technologies contribute to the construction of a greener future.

REFERENCES

[1] Javaid, M., Haleem, A., Singh, R. P., Suman, R., & Gonzalez, E. S. (2022). Understanding the adoption of Industry 4.0 technologies in improving environmental sustainability. *Sustainable Operations and Computers*, 3, 203–217.

[2] Oláh, J., Aburumman, N., Popp, J., Khan, M. A., Haddad, H., & Kitukutha, N. (2020). Impact of Industry 4.0 on environmental sustainability. *Sustainability*, 12(11), 4674.

[3] Kumar, S. S., Varol, T., Canai, A., Kumaran, S. T., & Uthayakumar, M. (2021). A review on the performance of the materials by surface modification through EDM. *International Journal of Lightweight Materials and Manufacture*, 4(1), 127–144.

[4] Singh, N. K., Singh, Y., Sharma, A., Singla, A., & Negi, P. (2021). An environmental-friendly electrical discharge machining using different sustainable techniques: a review. *Advances in Materials and Processing Technologies*, 7(4), 537–566.

[5] Ming, W., Cao, C., Xie, Z., Liu, X., Xu, Y., Jiang, Z., … & Li, L. (2022). Green manufacturing: a comparative study of renewable dielectrics in the EDM process. *Journal of the Brazilian Society of Mechanical Sciences and Engineering*, 44(12), 580.

[6] Arif, U., Ali Khan, I., & Hasan, F. (2023). Green and sustainable electric discharge machining: a review. *Advances in Materials and Processing Technologies*, 9(3), 970–1044.

[7] Ming, W., Xie, Z., Ma, J., Du, J., Zhang, G., Cao, C., & Zhang, Y. (2021). Critical review on sustainable techniques in electrical discharge machining. *Journal of Manufacturing Processes*, 72, 375–399.

[8] Dhakar, K., Kumar, R., Katheria, A., Nagdeve, L., & Kumar, H. (2022). Effect of various dielectric fluids on electric discharge machining (EDM): a review. *Journal of the Brazilian Society of Mechanical Sciences and Engineering*, 44(10), 487.

[9] Ming, W., Shen, F., Zhang, G., Liu, G., Du, J., & Chen, Z. (2021). Green machining: a framework for optimization of cutting parameters to minimize energy consumption and exhaust emissions during electrical discharge machining of Al 6061 and SKD 11. *Journal of Cleaner Production*, 285, 124889.

[10] Yang, M., Zhang, D., Wu, B., & Zhang, Y. (2020). Energy consumption modeling for EDM based on material removal rate. *IEEE Access*, 8, 173267–173275.

[11] Bajaj, R., Dixit, A. R., & Tiwari, A. K. (2020). Machining performance enhancement of powder mixed electric discharge machining using green dielectric fluid. *Journal of the Brazilian Society of Mechanical Sciences and Engineering*, 42, 1–20.

[12] Baroi, B. K., Jagadish, & Patowari, P. K. (2022). A review on sustainability, health, and safety issues of electrical discharge machining. *Journal of the Brazilian Society of Mechanical Sciences and Engineering*, 44(2), 59.

[13] Qudeiri, J. E. A., Zaiout, A., Mourad, A. H. I., Abidi, M. H., & Elkaseer, A. (2020). Principles and characteristics of different EDM processes in machining tool and die steels. *Applied Sciences*, 10(6), 2082.

[14] Ming, W., Zhang, Z., Wang, S., Zhang, Y., Shen, F., & Zhang, G. (2019). Comparative study of energy efficiency and environmental impact in magnetic field assisted and conventional electrical discharge machining. *Journal of Cleaner Production*, 214, 12–28.

[15] Zeid, O. A. (1997). On the effect of electrodischarge machining parameters on the fatigue life of AISI D6 tool steel. *Journal of Materials Processing Technology*, 68(1), 27–32.

[16] Jahan, M. P., Wong, Y. S., & Rahman, M. (2009). A study on the fine-finish die-sinking micro-EDM of tungsten carbide using different electrode materials. *Journal of Materials Processing Technology*, 209(8), 3956–3967.

[17] Yeo, S. H., Tan, P. C., Aligiri, E., Tor, S. B., & Loh, N. H. (2009). Processing of zirconium-based bulk metallic glass (BMG) using micro electrical discharge machining (Micro-EDM). *Materials and Manufacturing Processes*, 24(12), 1242–1248.

[18] Dhakar, K., & Dvivedi, A. (2017). Experimental investigation on near-dry EDM using glycerin-air mixture as dielectric medium. *Materials Today: Proceedings*, 4(4), 5344–5350.

[19] Mishra, B. P., & Routara, B. C. (2020). Evaluation of technical feasibility and environmental impact of Calophyllum Inophyllum (Polanga) oil based bio-dielectric fluid for green EDM. *Measurement*, 159, 107744.

[20] Valaki, J. B., Rathod, P. P., Khatri, B. C., & Vaghela, J. R. (2016, October). Investigations on palm oil based biodielectric fluid for sustainable electric discharge machining. In *Proceedings of the International Conference on Advances in Materials and Manufacturing* (ICAMM-2016), Bangkok, Thailand (pp. 29–30).

[21] Wu, X., Liu, Y., Zhang, X., Dong, H., Zheng, C., Zhang, F., ... & Ji, R. (2020). Sustainable and high-efficiency green electrical discharge machining milling method. *Journal of Cleaner Production*, 274, 123040.

[22] Pandey, A. K., & Anas, M. (2022). Sustainability and recent trends in micro-electric discharge machining (μ-EDM): a state-of-the-art review. *Materials Today: Proceedings*, 57, 2049–2055.

[23] Zhong, Z. W. (2021). Processes for environmentally friendly and/or cost-effective manufacturing. *Materials and Manufacturing Processes*, 36(9), 987–1009.

[24] Menzel, V., Smagin, J., & David, F. (2010). Can companies profit from greener manufacturing? *Measuring Business Excellence*, 14(2), 22–31.

[25] Pragadish, N., Kaliappan, S., Subramanian, M., Natrayan, L., Satish Prakash, K., Subbiah, R., & Kumar, T. C. A. (2023). Optimization of cardanol oil dielectric-activated EDM process parameters in machining of silicon steel. *Biomass Conversion and Biorefinery*, 13(15), 14087–14096.

[26] Rakshaskar, R., & Kannan, C. (2024). Experimental investigation and optimization with vegetable oil-based biodiesels as dielectric in EDM: green initiative towards sustainable manufacturing. *Journal of the Brazilian Society of Mechanical Sciences and Engineering*, 46(1), 57.

[27] Gugulothu, B. (2020). Optimization of process parameters on EDM of titanium alloy. *Materials Today: Proceedings*, 27, 257–262.

[28] Khan, F., & Soota, T. (2020). Optimization of EDM process parameter for stainless steel D3. *Materials Today: Proceedings*, 25, 635–638.

[29] Alrubaye, I. D. K., & Fantoni, G. (2023). Toward green electrical discharge machining (EDM): state of art and outlook. *Machining Science and Technology*, 27(1), 63–105.

[30] Uhlmann, E., Polte, M., & Yabroudi, S. (2022). Novel advances in machine tools, tool electrodes and processes for high-performance and high-precision EDM. *Procedia CIRP*, 113, 611–635.

[31] Gouda, D., Panda, A., Nanda, B. K., Kumar, R., Sahoo, A. K., & Routara, B. C. (2021). Recently evaluated electrical discharge machining (EDM) process performances: a research perspective. *Materials Today: Proceedings*, 44, 2087–2092.

[32] Schefer, H., Fauth, L., Kopp, T. H., Mallwitz, R., Friebe, J., & Kurrat, M. (2020). Discussion on electric power supply systems for all electric aircraft. *IEEE Access*, 8, 84188–84216.

[33] Zhao, W., Chen, M., Xia, W., Xi, X., Zhao, F., & Zhang, Y. (2020). Reconstructing CNC platform for EDM machines towards smart manufacturing. *Procedia CIRP*, 95, 161–177.

[34] Almeida, S. T., Mo, J. P., Bil, C., Ding, S., & Wang, X. (2022). Servo control strategies for vibration-control in robotic wire EDM machining. *Archives of Computational Methods in Engineering*, 29(1), 1–15.

[35] Ali, P., Walia, R. S., Murtaza, Q., & Singari, R. M. (2020). Material removal analysis of hybrid EDM-assisted centrifugal abrasive flow machining process for performance enhancement. *Journal of the Brazilian Society of Mechanical Sciences and Engineering*, 42, 1–28.

[36] Wu, Y. Y., Huang, T. W., & Sheu, D. Y. (2020). Desktop micro-EDM system for high-aspect ratio micro-hole drilling in tungsten cemented carbide by cut-side micro-tool. *Micromachines*, 11(7), 675.

[37] Singh, A., Ghadai, R. K., Kalita, K., Chatterjee, P., & Pamučar, D. (2020). EDM process parameter optimization for efficient machining of Inconel-718. *Facta Universitatis, Series: Mechanical Engineering*, 18(3), 473–490.

[38] Boopathi, S. (2023). Performance improvement of eco-friendly near-dry wire-cut electrical discharge machining process using coconut oil-mist dielectric fluid. *Journal of Advanced Manufacturing Systems,* 22(02), 339–358.

[39] Das, S., & Barma, J. D. (2022). Eco-friendly dielectric mediums for sustainable EDM: a comprehensive review. *Academic Journal of Manufacturing Engineering*, 20(4), 31–42.

[40] Baroi, B. K., Jagadish, & Patowari, P. K. (2022). A review on sustainability, health, and safety issues of electrical discharge machining. *Journal of the Brazilian Society of Mechanical Sciences and Engineering*, 44(2), 59.

[41] Suthar, J., Teli, S. N., & Katke, G. S. (2021). EDM parameters optimization for sustainable machining. *Materials Today: Proceedings*, 47, 2820–2824.

[42] Ming, W., Shen, F., Zhang, Z., Huang, H., Du, J., & Wu, J. (2020). A comparative investigation on magnetic field–assisted EDM of magnetic and non-magnetic materials. *The International Journal of Advanced Manufacturing Technology*, 109, 1103–1116.

[43] Nagabhooshanam, N., Baskar, S., Anitha, K., & Arumugam, S. (2021). Sustainable machining of hastelloy in EDM using nanoparticle-infused biodegradable dielectric fluid. *Arabian Journal for Science and Engineering*, 46(12), 11759–11770.

[44] Li, W., Liu, Q., Zhang, Y., Li, C. A., He, Z., Choy, W. C., ... & Kyaw, A. K. K. (2020). Biodegradable materials and green processing for green electronics. *Advanced Materials*, 32(33), 2001591.
[45] Grainger, A., & Smith, G. (2021). The role of low carbon and high carbon materials in carbon neutrality science and carbon economics. *Current Opinion in Environmental Sustainability*, 49, 164–189.
[46] Sizirici, B., Fseha, Y., Cho, C. S., Yildiz, I., & Byon, Y. J. (2021). A review of carbon footprint reduction in construction industry, from design to operation. *Materials,* 14(20), 6094.
[47] Wickramasinghe, K. C., Sasahara, H., Abd Rahim, E., & Perera, G. I. P. (2020). Green metalworking fluids for sustainable machining applications: a review. *Journal of Cleaner Production,* 257, 120552.
[48] Silva, L. R., Corrêa, E. C., Brandao, J. R., & de Avila, R. F. (2020). Environmentally friendly manufacturing: behavior analysis of minimum quantity of lubricant-MQL in grinding process. *Journal of Cleaner Production*, 256, 103287.
[49] Jiang, Y., Kong, L., Yu, J., Hua, C., & Zhao, W. (2022). Experimental investigation on the effects of green dielectric medium air and argon on the near-dry electrical discharge machining process. *Measurement,*120(11), 124889.
[50] Gupta, V., Singh, B., & Mishra, R. K. (2020). Machining of titanium and titanium alloys by electric discharge machining process: a review. *International Journal of Machining and Machinability of Materials*, 22(2), 99–121.
[51] Diviya, M., Joel, J. J., Subramanian, M., Balasubramanian, T., Madhusuthan, A. V., Monish, N., & Hasan, N. (2024). Parametric investigation of W-EDM factors for machining AM60B conductive biomaterial. *Scientific Reports,* 14(1), 216.
[52] Das, S., Paul, S., & Doloi, B. (2019). An experimental and computational study on the feasibility of bio-dielectrics for sustainable electrical discharge machining. *Journal of Manufacturing Processes*, 41, 284–296.
[53] Singh, N. K., Yadav, L., & Lal, S. (2022). Experimental investigation for sustainable electric discharge machining with Pongamia and Jatropha as dielectric medium. *Advances in Materials and Processing Technologies*, 8(2), 1447–1466.
[54] Bains, P. S., Singh, S., Sidhu, S. S., Kaur, S., & Ablyaz, T. R. (2018). Investigation of surface properties of Al–SiC composites in hybrid electrical discharge machining. *Futuristic Composites: Behavior, Characterization, and Manufacturing*, 1, 181–196.
[55] Klocke, F., Zeis, M., Klink, A., & Veselovac, D. (2012). Technological and economical comparison of roughing strategies via milling, EDM and ECM for titanium-and nickel-based blisks. *Procedia CIRP*, 2, 98–101.
[56] Hourmand, M., Sarhan, A. A., & Sayuti, M. (2017). Micro-electrode fabrication processes for micro-EDM drilling and milling: a state-of-the-art review. *The International Journal of Advanced Manufacturing Technology,* 91(1), 1023–1056.
[57] Kou, Z., & Han, F. (2018). On sustainable manufacturing titanium alloy by high-speed EDM milling with moving electric arcs while using water-based dielectric. *Journal of Cleaner Production*, 189, 78–87.
[58] Prabhu, S., & Vinayagam, B. K. (2010). Analysis of surface characteristics of AISI D2 tool steel material using electric discharge machining process with single wall carbon nano tubes. *IACSIT International Journal of Engineering and Technology*, 2(1), 1793–8236.
[59] De Jesus, A. M. P., Ramos, G. F. S., Gomes, V. M. G., Marques, M. J., De Figueiredo, M. A. V., & Marafona, J. D. R. (2020). Comparison between EDM and grinding machining on fatigue behaviour of AISI D2 tool steel. *International Journal of Fatigue*, 139(105742), 1–11.

[60] Pragadish, N., & Kumar, M. P. (2015). Surface characteristics analysis of dry EDMed AISI D2 steel using modified tool design. *Journal of Mechanical Science and Technology,* 29(4), 1737–1743.

[61] Abbas, M. A., Lajis, M. A., Abbas, D. R., Merzah, O. M., Kadhim, M. H., & Shamran, A. A. (2020). Influence of additive materials on the roughness of AISI D2 steel in electrical discharge machining (EDM) environment. *Materwiss Werksttech,* 51(6), 719–724.

[62] Khan, F., Kumar, J., & Soota, T. (2019). Optimization of EDM process parameter for stainless steel D3. *Materials Today: Proceedings,* 25, 635–638.

[63] Hamid, F. E. A., & Lajis, M. A. (2012). High performance in EDM machining of AISI D2 hardened steel. *Advanced Materials Research,* 500, 259–265.

[64] Singh, N., Routara, B. C., & Nayak, R. K. (2018). Study of machining characteristics of Inconel 601 with cryogenic cooled electrode in EDM using RSM. *Materials Today: Proceedings,* 5(11), 24277–24286.

[65] Prakash, D., Tariq, M., Davis, R., Singh, A., & Debnath, K. (2021). Influence of cryogenic treatment on the performance of micro-EDM tool electrode in machining of magnesium alloy AZ31B. *Materials Today: Proceedings,* 39, 1198–1201.

[66] Anand, B., Giri, A., Mohanty, C. P., & Sharma, D. (2020). Tool wear and energy consumption optimization in EDM of chromium tool steel. *Materials Today: Proceedings,* 43, 268–272.

[67] Rao, K. M., Kumar, D. V., Shekar, K. C., & Singaravel, B. (2021). Optimization of EDM process parameters using TOPSIS for machining AISI D2 steel material. *Materials Today: Proceedings,* 46, 701–706.

[68] Gopalakannan, S., Senthilvelan, T., & Kalaichelvan, K. (2012). Electrical discharge machining of hybrid metal matrix composites by applying Taguchi method. *International Journal of Manufacturing Technology and Management,* 26(1–4), 114–136.

2 Advancements in Sustainability in Electrical Discharge Machining with Bio-Dielectric Fluids

Sameer Kumar Anand and Dupadu Chakradhar

ABBREVIATIONS

BDF	Bio-dielectric fluid
BMG	Bulk metallic glass
EDM	Electrical discharge machining
HBDF	Hydrocarbon-based dielectric fluid
MRR	Material removal rate
PBD	Polanga bio-dielectric
PMEDM	Powder-mixed EDM
SEM	Scanning electron microscope
SR	Surface roughness
TWR	Tool wear rate
VOBDF	Vegetable oil-based dielectric fluid
WVO	Waste vegetable oil

2.1 INTRODUCTION

Electrical Discharge Machining (EDM) is an unconventional subtracting manufacturing technique that is becoming a pivotal manufacturing technique, particularly renowned for its effectiveness in dealing with the intricacies of hard-to-cut materials with exceptional precision. This technique has been indispensable in the realm of precision manufacturing for years. By employing high-frequency electrical discharges, the dielectric fluid is thermoelectrically ionized in the EDM. Because of this, the temperature is elevated close to 8,000–12,000 K by the plasma produced during the technique, leading to the melting followed by vaporization of the work material. The versatility of EDM is evident across a wide range of industries, where it is utilized for machining aircraft components, automotive parts, biomedical devices [1], and various high-temperature application components. It is well-suited for materials, including hardened steel, titanium, and exotic alloys, that pose challenges for conventional machining techniques. In the EDM technique, the dielectric fluid is a

DOI: 10.1201/9781003769675-2

crucial element that performs several functions throughout the machining process. To facilitate stable and controlled machining, the dielectric fluid functions as the discharge fluid. In addition, these dielectric fluids serve to dissipate heat energy through electric sparks, which prevent overheating and potential damage to the workpiece. Moreover, dielectric fluid assists in effectively removing debris and byproducts generated during machining, maintaining cleanliness in the work environment to prevent potential clogging or obstruction in the technique.

The employment of conventional or hydrocarbon-based dielectric fluids (HBDFs) in EDM techniques raises environmental and health implications. Conventional dielectric fluids, such as transformer oil, kerosene, or mineral oil-based dielectrics, carry significant risks because of their toxicity and the potential for environmental contamination. Meticulously disposing of these fluids is essential to prevent soil and water pollution. In contrast, bio-dielectric fluids (BDFs) provide a potential solution for more sustainable EDM practices by offering biodegradability and non-toxic properties. Furthermore, BDFs contribute to consistently meeting industrial fluid regulations, ensuring compliance and minimizing penalties. Their adoption may signify a commitment to environmental and social responsibility, contributing to broader sustainability initiatives. Hence, a significant shift towards the exploration of BDFs in EDM is an alternative potential way towards sustainability in the EDM technique. The adoption of BDFs facilitates a sustainable alternative to mitigate environmental damage caused by soil pollution, water pollution, and emissions of volatile organic compounds. In addition, BDFs like vegetable oils or esters contribute to the conservation of natural resources by lessening dependence on non-renewable alternatives. In addition, vegetable oils are considered potential BDFs due to their higher oxidative stability and the presence of higher oleic acid in fatty acids [2]. Vegetable oil-based dielectric fluids (VOBDFs) are quickly biodegradable and less or non-toxic in nature, in contrast to the mineral oil-based dielectric fluids utilized in EDM practices [3], which not only improve environmental sustainability but also mitigate occupational health hazards and workplace safety risks. This is primarily due to the absence of polynuclear aromatics, halogens, and other potentially harmful compounds typically found in mineral fluids. The better biodegradation of vegetable oils is observed in both aerobic and anaerobic environments compared to mineral oil. The properties of VOBDFs have been illustrated in Figure 2.1. Furthermore, opting for BDFs aligns with regulatory requirements that ensure adherence to environmental norms and mitigate legal risks for manufacturing facilities. It has been observed that the incorporation of sustainable practices boosts their reputation, attracts environmentally conscious consumers, and promotes long-term sustainability in the manufacturing sector. Additionally, BDFs provide economic advantages by establishing markets for renewable resources, encouraging innovation, and advancing growth in the green technology sector. Thus, incorporating BDFs into EDM addresses environmental concerns, conserves natural resources, promotes economic viability, enhances worker health and safety, and meets regulatory standards. Hence, the advancements in sustainability within EDM are becoming critical as industries aim to manage economic progress with environmental and societal considerations in the competitive market.

In this chapter, the progress of sustainability in EDM has been discussed, with particular consideration given to the incorporation of BDFs as a dielectric medium.

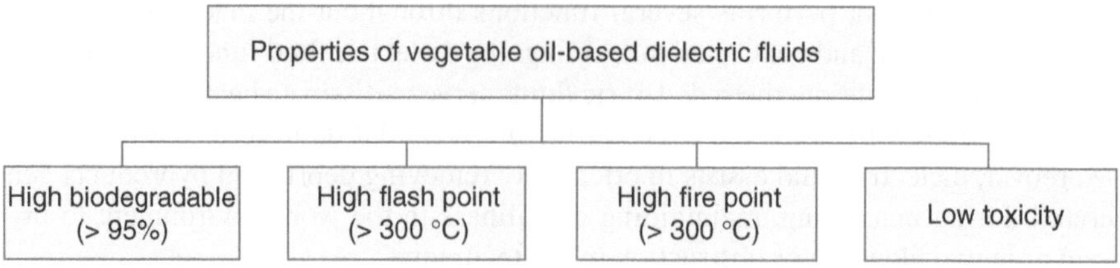

FIGURE 2.1 The properties of dielectric fluids based on vegetable oil [4].

It outlines the challenges posed by conventional dielectric fluids and explores how bio-dielectric-based alternatives offer promising approaches for enhancing sustainability in the EDM process. Furthermore, this chapter showcases emerging techniques and advancements that contribute to the sustainable evolution of the EDM technique, offering a glimpse into future challenges and opportunities within this domain.

2.2 SUSTAINABILITY CHALLENGES AND APPROACHES IN EDM

EDM was invented in the late 1940s and offered a unique solution to machine complex features with exceptional precision and improved surface qualities. From the late 1940s to the present day, EDM has experienced significant advancements. Thus, EDM facilitates several advantages over both conventional and unconventional machining methods, and therefore, nowadays, EDM is becoming indispensable in various sectors like automotive, defence and military, aerospace, medical, and tool and die-making [5,6]. Despite numerous advantages, EDM has limitations that inhibit its adoption in specific applications and industries. The EDM technique raises serious concerns for personnel health, as it releases toxic gases and fumes, micro-/nano-particles, odours, and volatile agents [7–9]. The major limitations of EDM include high initial setup costs, the necessity of material conductivity, slow machining speeds, surface characteristic issues, cost efficiency, social viability, and environmental implications. In recent decades, researchers have dedicated significant efforts to address challenges such as slow machining speeds and surface characteristic issues in EDM. However, the sustainability aspects of EDM have received relatively limited focus, making it a lively and growing research domain. Sustainability considers social, economic, and environmental aspects to safeguard resources, maintain ecological balance, promote social equality, and ensure long-term economic viability. Sustainable manufacturing aims to enhance environmental performance, improve energy efficiency, and minimize waste generation while prioritizing personal health and operational safety [10]. The evaluation of the manufacturing process's sustainability is assessed, as illustrated in Figure 2.2.

2.3 CONVENTIONAL DIELECTRIC FLUID: SUSTAINABILITY CONCERNS AND IMPLICATIONS

EDM is known for its exceptional precision and efficiency in the realm of material removal techniques. However, the utilization of conventional dielectric fluids

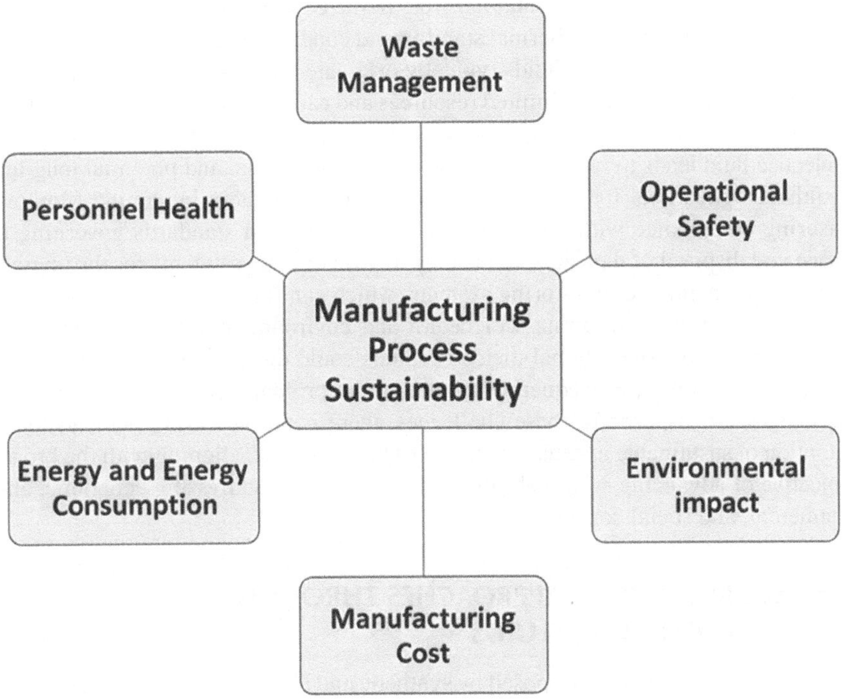

FIGURE 2.2 Criteria for sustainability of manufacturing process [11].

like synthetic and HBDFs raises sustainability concerns in EDM techniques. These sustainability concerns are primarily observed concerning their environmental, economic, and social implications. The major sustainability concerns linked to the utilization of synthetic and HBDFs include:

(a) Harmful emissions, including vapours and aerosols
(b) Carcinogenic emissions like polycyclic aromatic hydrocarbons, benzene, carbon monoxide, and carbon dioxide
(c) Generation of non-biodegradable waste products
(d) Waste disposal issues
(e) Regulatory compliance issues
(f) Personnel health issues
(g) Poor operational safety
(h) High specific energy consumption
(i) Cost ineffective

These sustainability concerns are raised by various factors, such as the non-biodegradability of dielectric fluid, the energy-intensive nature of the technique, and its potential to impact air, land, and water pollution. The production and disposal of synthetic dielectric fluids worsen climate change through greenhouse gas emissions. Moreover, improper handling of the chemical composition of these dielectric fluids leads to health and environmental hazards. In the case of HBDFs, they offer advantages in terms of improved thermal stability and conductivity, but they also raise sustainability concerns. These fluids typically originate from petroleum or mineral oil, which raises the demand for limited resources and causes environmental degradation during extraction and refining processes. Exposure to emissions of conventional dielectric fluid leads to respiratory problems, skin irritation, and potential long-term health complications for personnel employed in manufacturing units. Moreover, ensuring compliance with environmental regulations and standards governing the usage and disposal of dielectric fluids may require high investments for the upgradation of equipment and monitoring systems, which can impact operational expenses. In addition, growing awareness of health and environmental concerns associated with the usage of conventional dielectric fluids could drive market demand for sustainable alternatives, consequently impacting the profitability and sustainability of businesses. In response to these challenges, there is an increasing demand for the adoption of sustainable alternatives in the EDM technique, aligning with the broader objective of advancing sustainability across sectors and addressing economic, environmental, and social concerns.

2.4 SUSTAINABILITY APPROACHES THROUGH BIO-DIELECTRIC FLUIDS

The sustainability challenges posed by synthetic and HBDFs suggest an alternative to the adoption of BDF in the EDM technique. BDFs, derived from renewable sources such as vegetable oils, offer an alternative option. BDFs offer various advantages over synthetic and hydrocarbon-based fluids in the EDM technique, which include:

(a) Environmental friendliness
(b) Biodegradable
(c) Less toxic/non-toxic
(d) Sustainable waste management
(e) Compliance with regulatory standards
(f) Improved personnel health
(g) Improved operational safety
(h) Economically viable
(i) Improved machining performances

BDFs, extracted from renewable sources, emerge as a promising substitute for synthetic and HBDFs in the EDM technique. The biodegradability of BDFs facilitates their natural breakdown into harmless components over time, thereby lowering the risk of potential soil and water pollution. In addition, BDFs help to lower carbon emissions and greenhouse gases and lessen our dependence on limited fossil fuel reserves. Moreover, the non-toxic nature of BDFs used in EDM enhances safety for workers and provides a better working environment. These factors ensure adherence to regulations and reduce environmental pollution to make it more eco-friendly. Apart from this, the utilization of BDFs in the EDM technique contributes to the well-being of communities. Even though BDFs may have a higher or comparable cost to conventional dielectric fluids, the potential for lower disposal expenses, reuse opportunities, and reduced regulatory compliance costs could result in significant overall savings. There is a promising prospect for BDFs in EDM, as ongoing research and development endeavours target to enhance their performance and sustainability even further. The advancement of technology, particularly in the creation of innovative BDFs and additives to improve conductivity and machining efficiency, holds potential for further advancements in this area.

2.5 TECHNIQUES AND INNOVATIONS FOR SUSTAINABLE EDM

The EDM technique has remained significant in the realm of manufacturing over time and is providing a precise and effective approach to the machining of conductive materials. However, similar to several conventional manufacturing methods, EDM has encountered criticism related to sustainability. In response to these concerns, researchers have been actively exploring different methods and innovations aimed at enhancing the sustainability of EDM, as illustrated in Figure 2.3.

Dry EDM, often referred to as near-dry EDM, is an innovative machining technique that also contributes to the sustainability of EDM techniques. In contrast to conventional EDM techniques that rely on conventional dielectric fluids to facilitate spark erosion, dry EDM operates without the need for these conventional dielectric fluids. Comparable or better results are achieved by employing controlled atmospheres of inert gases such as nitrogen or argon [12]. In addition, dry EDM decreases the environmental impact as well as the operational costs associated with the disposal and recycling of the dielectric fluid [13]. However, adopting the dry EDM technique requires sophisticated equipment that can sustain precise inert gas atmospheres. Process optimization is also essential, which involves adjustments to factors like gas composition and pressure to ensure efficient spark generation and debris

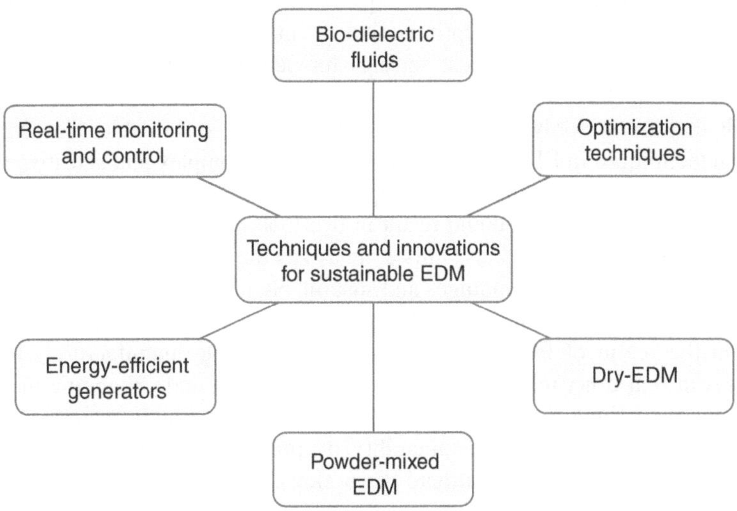

FIGURE 2.3 Major techniques and innovations for sustainable EDM.

removal. It is vital to optimize processes by adjusting gas composition and pressure to maximize spark efficiency and debris removal. Moreover, the efficient and safe operation of the dry EDM technique is ensured by providing proper training for the operator. In addition, operators must have a thorough understanding of the specific aspects of these techniques and be skilled in handling any potential challenges that may arise during the operation.

The powder-mixed EDM (PMEDM) technique has also revolutionized conventional EDM by incorporating electrically conductive powders in the dielectric fluid. The incorporation of powder enhances not only the machining efficiency and precision but also the sustainability of the EDM techniques. By minimizing both energy consumption and waste generation, the focus of PMEDM on sustainability makes it remarkable in environmentally responsible manufacturing. Despite several significant advantages, there are various challenges, including increased powder complexities, powder recycling and reusing, uniform powder dispersion, material removal uniformity, and health and safety concerns associated with the operation of the PMEDM technique [14].

Apart from these techniques, modern EDM machines feature energy-efficient generators that optimize power usage while maintaining high machining performance [14]. These generators include advanced power electronics and control algorithms to regulate energy consumption during machining, resulting in reduced operating expenses and environmental impact. The synchronization of efficiency and precision facilitates manufacturers to achieve significant cost reductions without compromising productivity, demonstrating a broader commitment to sustainable and innovative manufacturing techniques. However, modern EDM machines featuring energy-efficient generators require specialized training for operators as well as maintenance personnel due to their complexity. Furthermore, the high initial investment required to acquire and install these sophisticated machines may result in financial challenges for manufacturers.

Optimization is also one of the techniques being implemented to minimize material waste and energy usage. Optimization techniques analyse a range of control parameters to determine the best settings for achieving optimal machining performance and surface integrity. This approach fosters the adoption of machining methods that are not only more efficient but also more sustainable. However, optimization techniques in EDM can be computationally complex and sensitive to input parameters, leading to challenges in implementation. Moreover, there is a chance of over-optimization, which could result in overlooking critical factors such as tool wear and surface characteristics. This emphasizes the importance of finding a balance between short-term advantages and sustainable machining practices.

Real-time monitoring and control systems also represent substantial advancements in the realm of EDM [15]. These systems enable manufacturers to adjust processes in real time, mitigating resource consumption and enhancing machining efficiency. In addition to improving machining efficiency and resource consumption, these systems promote sustainability by proactively maintaining equipment to prevent downtime and operational disruptions, managing operations to reduce emissions and environmental impact, and supporting decision-makers by analysing data. With the utilization of real-time data analysis, improvement in the operational efficiency of the system is focused, ultimately driving towards a more sustainable EDM technique. Despite several benefits, initial investment costs, the possibility of system failures, and complexities demanding specific skills are major concerns about real-time monitoring in EDM techniques.

In the pursuit of enhancing the sustainability of the EDM technique, vegetable oil-based fluids have emerged as potential BDFs, offering a viable solution to the concerns associated with EDM [16,17]. Vegetable oils originate from renewable sources, thereby decreasing the dependency on limited fossil fuels and reducing carbon emissions. Vegetable oils are considered potential BDFs due to their higher oxidative stability and the presence of higher oleic acid in fatty acids [2]. The dielectric fluids' desired properties have been illustrated in Table 2.1. The utilization of VOBDFs exhibits a superior sustainability index for industrial applications when contrasted with synthetic and hydrocarbon-based fluids [16]. Furthermore, as a consequence of higher flash points, BDFs represent fewer health concerns for workers and present fewer fire hazards. Moreover, they can improve machining performance by offering a higher material removal rate (MRR) and better surface finishes. Thus, BDFs offer a superior alternative over conventional dielectric fluids in EDM to make it more sustainable. Advancements in the conventional EDM process due to the incorporation of BDFs have been discussed in the subsequent section.

2.6 ADVANCEMENTS IN THE CONVENTIONAL EDM PROCESS

In addition to enhancing established techniques, advancements in conventional EDM aim to incorporate sustainable practices, such as the use of BDFs. The BDFs obtained from renewable resources like vegetable oils exhibit promise for mitigating environmental impact while enhancing surface characteristics and machining efficiency. To satisfy the growing demands of modern industry, the conventional EDM method has experienced significant improvement with the employment of BDFs as dielectric fluids in recent years. Following is the research that has been conducted on advancements in the conventional EDM technique, with a focus on the utilization of BDFs as dielectric fluids.

Valaki et al. [18] introduced the use of palm oil as a dielectric medium in the EDM as a sustainable alternative to kerosene dielectric fluid. The outcomes indicated that EDM in palm oil resulted in a 38% higher MRR, with surface hardness and surface roughness (SR) remaining nearly unchanged, in contrast to hydrocarbon-based kerosene. Palm oil was recommended as an alternative to kerosene dielectric fluid in EDM, which is highly unsustainable in contrast to kerosene dielectric fluid. Valaki and Rathod [17] investigated the potential of the green dielectric (Jatropha oil and waste vegetable oil (WVO)) to improve sustainability in the EDM of M238 HH grade cold-worked mould steel. Findings demonstrated that the newly proposed bio-dielectrics, Jatropha oil and WVO, outperformed kerosene in both MRR and relative wear ratio. When subjected to electrical current, Jatropha oil and WVO demonstrated enhancements in MRR by 38% and 165%, respectively. Furthermore, when subjected to pulse on time, Jatropha oil exhibited a 30% increase, while WVO showed a 7% decrease in relative wear ratio.

TABLE 2.1
Desired properties of dielectric employed in EDM. [Reprinted with permission from [19]]

Dielectric property	Desired	Hydrocarbon oil or mineral oil	Kerosene	Water	Bio-dielectric
Specific gravity	Low	High	Moderate	Low	Low
Flash point	High	Low	Low	High	High
Oxygen content	High	Low	Low	High	High
Toxicity	Low	High	High	Low	Low
Breakdown voltage	High	Low	Low	Moderate	High
Viscosity	Low	High	Low	Low	Low
Volatility	High	High	High	Moderate	High
Carbon content	Low	High	High	Negligible	Negligible
Biodegradability	More	Low	Low	More	More

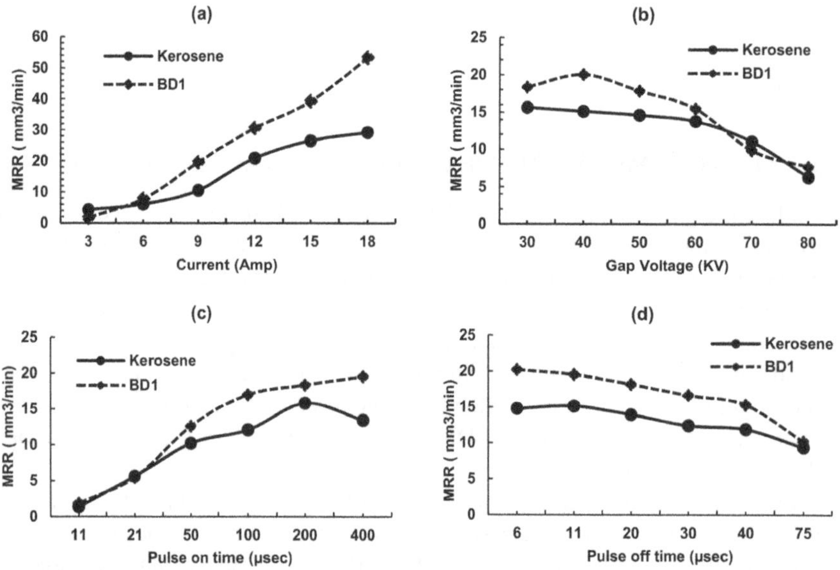

FIGURE 2.4 Comparison of the effects of control parameters on MRR for Jatropha curcas oil and kerosene. [Reprinted with permission from [20]]

Valaki et al. [20] examined the viability of a BDF, named Jatropha curcas oil, for EDM of plastic mould steel of M238 HH grade, with the goal of enhancing sustainability in the technique. The findings indicated that EDM performed in Jatropha bio-dielectric produced higher MRR, lower SR, and higher surface hardness in contrast to EDM performed in kerosene dielectric fluid. EDM in Jatropha bio-dielectric resulted in MRR values that were 15%, 28%, 26%, and 38% higher in contrast to EDM in kerosene across different process variables, including pulse off time, pulse on time, gap voltage, and current, respectively, as illustrated in Figure 2.4. Similarly,

EDM in Jatropha bio-dielectric resulted in SR values that were 6%, 18%, 13%, and 23% lower in contrast to EDM in kerosene across different process variables, including pulse off time, pulse on time, gap voltage, and current, respectively, as illustrated in Figure 2.5. In terms of surface hardness, EDM in Jatropha bio-dielectric showed values that were 5%, 6%, 7%, and 6% higher in contrast to EDM in kerosene across different process variables, including pulse off time, pulse on time, gap voltage, and current, respectively, as illustrated in Figure 2.6. Moreover, the response patterns exhibited in Jatropha bio-dielectric closely resembled those of kerosene and suggested that the mechanisms of material melting and vaporation are alike in both dielectric fluids. It was recommended that Jatropha oil could serve as a viable alternative to hydrocarbon dielectrics, thereby enhancing the sustainability of the EDM technique.

Valaki and Rathod [21] investigated the practicality of employing WVO as a BDF to make the EDM technique more sustainable. Comparisons of control parameters on MRR, electrode wear rate, and tool wear ratio for WVO and kerosene have been illustrated in Figures 2.7, 2.8, and 2.9, respectively. It was revealed through the investigation that the WVO-as dielectric fluid emerged as a promising option, potentially replacing synthetic, water, and HBDFs in the EDM. Besides confirmation of operational feasibility, it was found that the utilization of BDFs offers a cleaner, safer, and greener dielectric solution, aiming to enhance the sustainability of the EDM technique by addressing environmental, personnel health, and operational safety concerns. Ng et al. [22] explored the viability and sustainability of utilizing biodiesel as a dielectric fluid in the EDM of Ti-6Al-4V and bulk metallic glass (BMG) materials using a tool electrode of 200 μm. Sunflower biodiesel and canola biodiesel

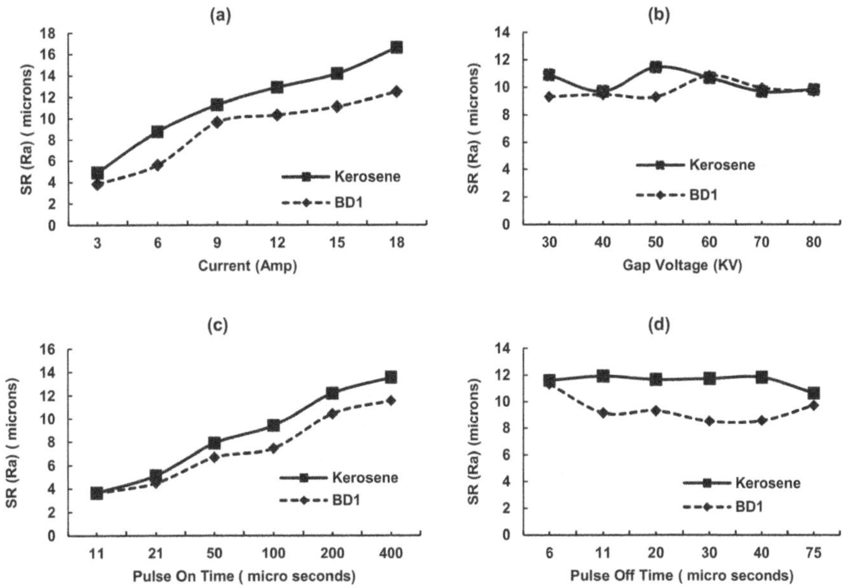

FIGURE 2.5 Comparison of the effects of control parameters on SR for Jatropha curcas oil and kerosene. [Reprinted with permission from [20]]

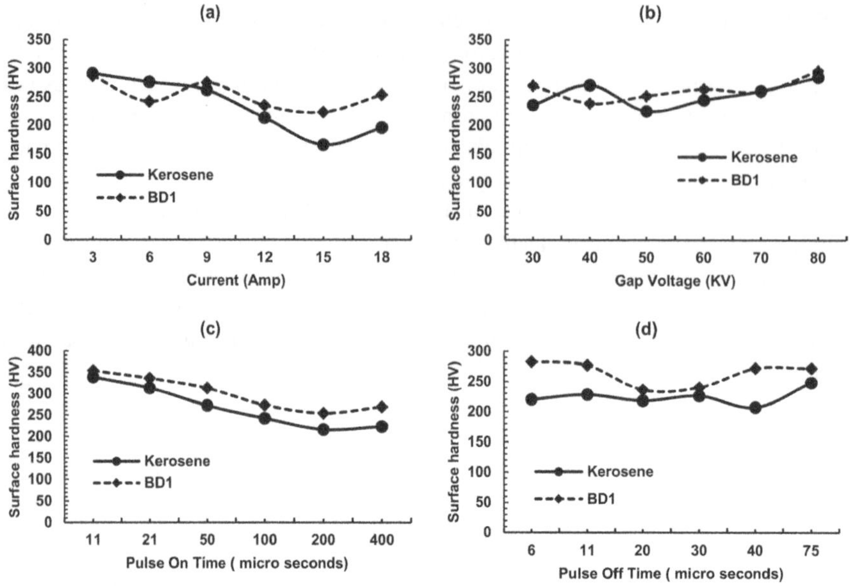

FIGURE 2.6 Comparison of the effects of control parameters on surface hardness for Jatropha curcas oil and kerosene. [Reprinted with permission from [20]]

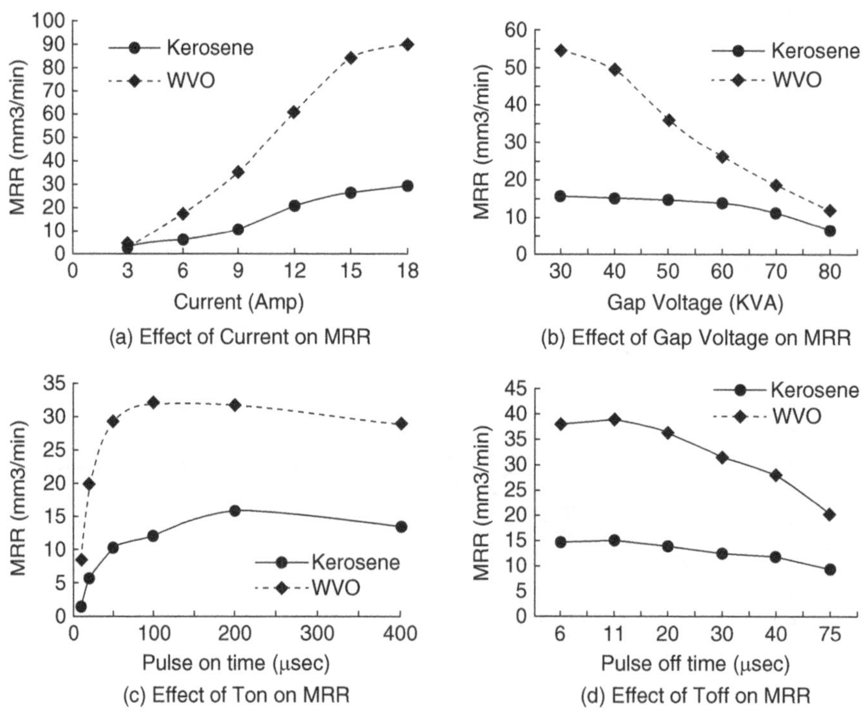

FIGURE 2.7 Comparison of control parameters on MRR for WVO and kerosene. [Reprinted with permission from [21]]

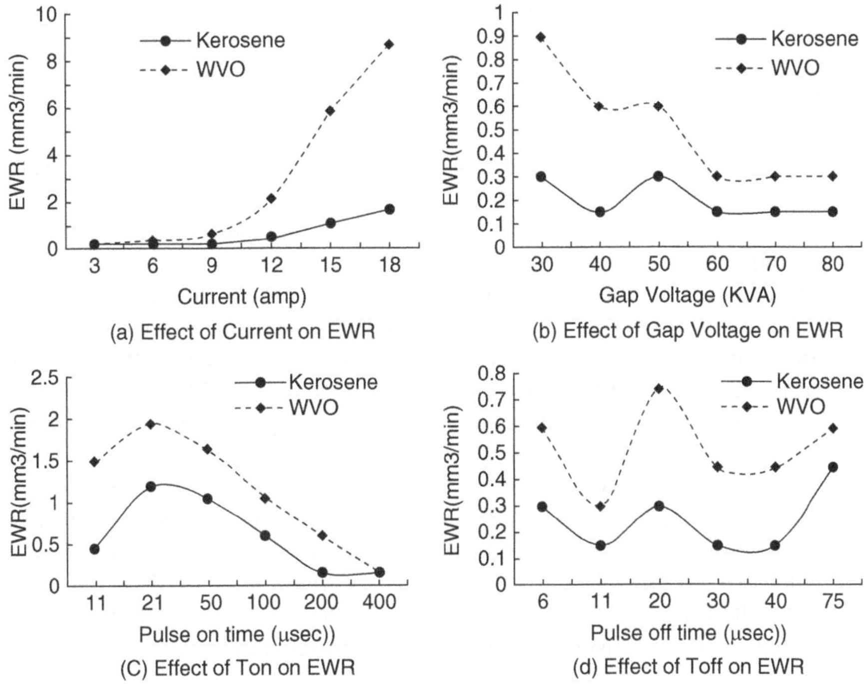

FIGURE 2.8 Comparison of control parameters on electrode wear rate for WVO and kerosene. [Reprinted with permission from [21]]

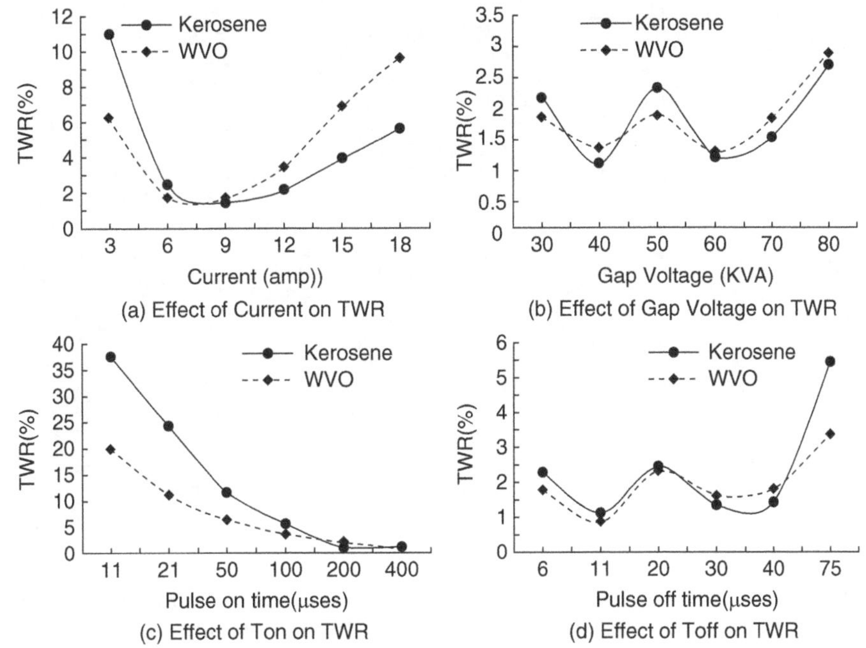

FIGURE 2.9 Comparison of control parameters on tool wear ratio for WVO and kerosene. [Reprinted with permission from [21]]

were chosen for this experiment owing to their high flash point, low viscosity, and fatty acid composition [23]. The experimental results indicated that both canola and sunflower biodiesel surpassed conventional dielectric fluids in performance. While machining BMG, the MRR using canola and sunflower biodiesel surpassed that of traditional dielectric fluid by 114% to 137%. Similarly, for the machining of Ti-6Al-4V, the MRR achieved with canola and sunflower biodiesel exceeded that of traditional dielectric fluid by 12% to 108%. These facts indicate the potential for biodiesel to replace traditional dielectrics, which may offer a more sustainable machining technique.

Sadagopan and Mouliprasanth [24] investigated the EDM of aluminium alloy 6063 specimens by utilizing three distinct dielectric fluids, including biodiesel, kerosene, and transformer oil. The findings showed that EDM in biodiesel, as a dielectric fluid, demonstrated superior performance in MRR, tool wear rate (TWR), and SR compared to EDM in kerosene and transformer oil. Das et al. [25] examined the potential of using a bio-dielectric derived from jatropha, canola, and neem as a substitute for conventional dielectric fluid, kerosene. The study revealed that all BDFs exhibited enhanced MRR and considerable SR compared to kerosene. However, jatropha and canola demonstrated particularly remarkable productivity in both MRR and SR. It was found that bio-dielectrics of this nature demonstrated ordinary MRR when operated at low flushing speeds. As a result, only a constrained range of flushing speeds is required to obtain satisfactory outcomes. Singaravel et al. [26,27] examined the

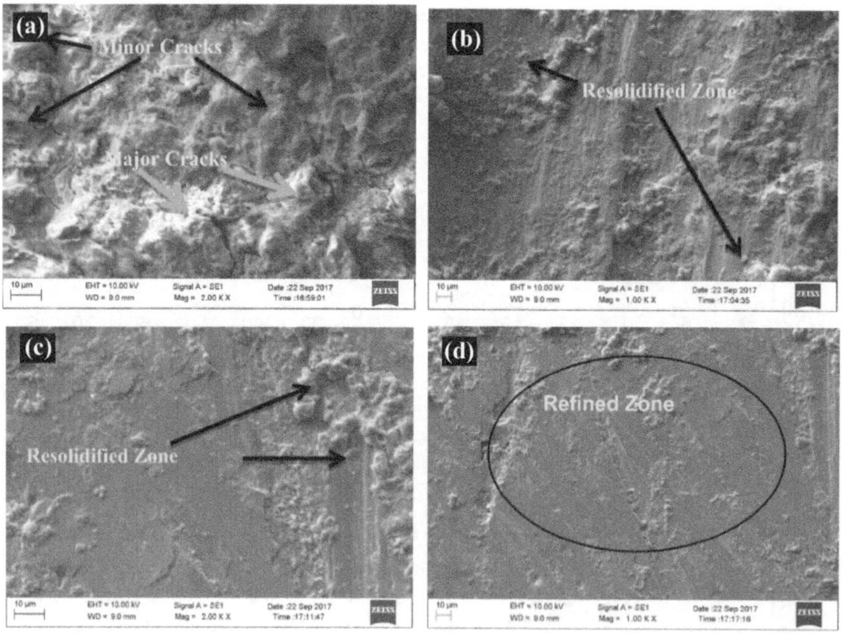

FIGURE 2.10 SEM micrograph of the workpiece machined with (a) hydrocarbon dielectric fluid at 2,000× magnification, (b) Polanga BDF at 2,000×, (c) hydrocarbon dielectric fluid at 1,000× magnification, and (d) Polanga BDF at 1000×. [Reprinted with permission from [28]].

impact of employing vegetable oils as dielectric fluids on the machining performance of titanium alloy and Inconel 800 in two distinct studies. During the investigation, the authors compared the impact of conventional dielectric fluids and vegetable oils on the efficiency of the EDM technique. The findings suggest that vegetable oils operate with a material removal mechanism similar to conventional dielectric fluids, thereby hinting at their possible substitution in EDM techniques.

Mishra and Routara [28] investigated the suitability of Polanga bio-dielectric (PBD) as a substitute bio-dielectric for EDM techniques and analysed its process response relative to EDM oil. The study findings showed that Polanga BDF outperformed conventional dielectric kerosene, resulting in increased productivity and enhanced microstructure performance. The scanning electron microscope (SEM) analysis showed improvements in microstructures, and SR achieved with the PBD medium in contrast to the traditional hydrocarbon dielectric medium. These distinct differences have been illustrated in Figure 2.10. The surfaces produced using hydrocarbon dielectric fluid exhibited noticeable undesirable issues, including damages, resolidified zones, minor cracks, and major cracks, all of which have been illustrated in Figure 2.10a and c. In the comparison of the utilization of hydrocarbon dielectric fluid, the utilization of Polanga as a BDF resulted in a confirmed reduction of 17.33% in aerosol emission rates, signifying a favourable environmental impact. The quantified rates of aerosol concentration corresponding to different control parameters have been illustrated in Figure 2.11. In the illustrated figure, hydrocarbon oil and PBD fluids have been abbreviated by HC and PBD, respectively. Moreover, the study

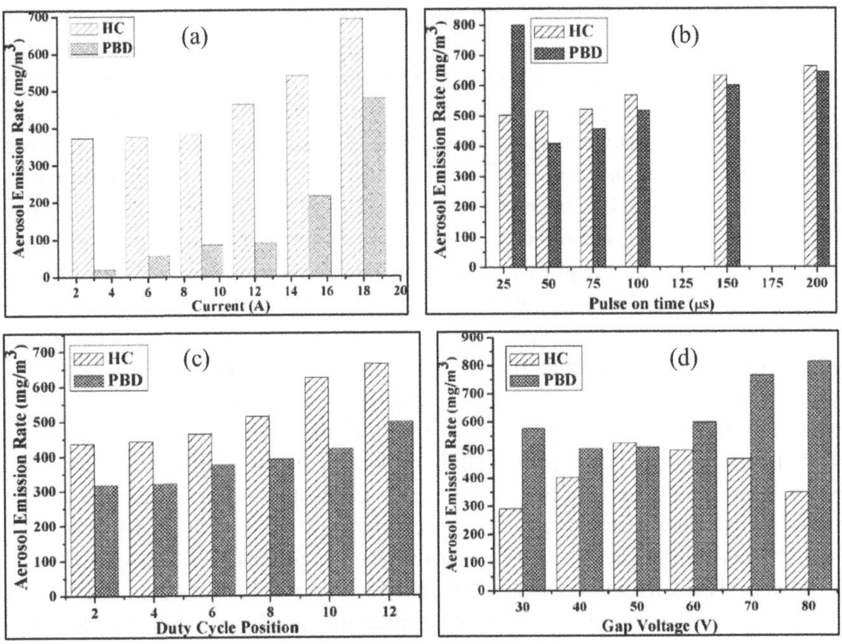

FIGURE 2.11 Comparative analysis of aerosol emission rate vs (a) peak current, (b) pulse on time, (c) duty cycle, and (d) gap voltage. [Reprinted with permission from [28]]

suggested its potential to contribute significantly to green manufacturing through the introduction of vegetable oil-based bio-dielectrics.

Singh et al. [2] explored the viability of employing Pongamia pinnata and Jatropha curcas as dielectric fluids for EDM of Al-SiC work material. The findings indicated that EDM in the proposed dielectric fluids performed better MRR, TWR, and SR than EDM using kerosene dielectric fluid. The EDM in Pongamia bio-dielectric showed 45%–75% higher MRR, while Jatropha bio-dielectric showed 30%–60% higher MRR compared to the EDM in kerosene dielectric fluid under similar control variables. Furthermore, EDM in the Jatropha and Pongamia bio-dielectrics led to roughly 18% and 23% lower TWR, respectively, compared to EDM in the conventional dielectric fluid. Moreover, machined specimens showed enhanced surface integrity. The findings also showed that the mechanics governing erosion and material removal processes closely resembled those observed with kerosene dielectric fluid when bio-dielectric was utilized. Ultimately, it was suggested that the proposed dielectric fluids, namely Jatropha oil and Pongamia, could be utilized as an alternative to HBDFs to enhance the EDM technique's sustainability.

2.7 CHALLENGES AND FUTURE PERSPECTIVES

Various research has been carried out to utilize diverse dielectric fluids to enhance the efficiency of EDM technique in terms of machining performance and surface integrity, thereby promoting sustainability. Furthermore, the scope of research can be broadened to include not only the investigation of the influence of different BDFs but also the exploration of the size and concentrations of powders or additives in the PMEDM technique that could help to analyse its effects on various response parameters for different workpiece specimens. These analyses could provide insights into determining the optimized condition for the PMEDM technique in different BDFs for various workpiece specimens. PMEDM has been proven to be an innovative technique to enhance the efficiency of EDM in hydrocarbon oil-based dielectric fluid. However, there has been a very limited exploration into PMEDM utilizing BDFs. Therefore, researchers can focus on this area to achieve promising results, emphasizing more sustainability aspects of the EDM technique. While BDFs are biodegradable, their biodegradability may be significantly affected by mixing them with powders or additives. Consequently, the discharge of such waste could pose significant environmental problems. Therefore, research should be focused on finding powders whose discharging should have fewer environmental concerns without compromising the surface quality and machining performance of workpiece specimens. Furthermore, there is no or limited study related to the durability of BDFs, resulting in conflicting evaluations of their suitability for repeated use. As a result, it becomes crucial to conduct research focused on estimating the lifespan of different BDFs used in EDM techniques for various workpiece specimens. The economic viability of BDFs is also a challenging task in the EDM technique. Even though they have sustainability benefits, their higher price compared to that of conventional dielectric fluids is preventing them from being widely used. Thus, further exploration is required to optimize production techniques and reduce expenses. It is also essential to ensure that BDFs are stable and compatible with various materials. The performance of

BDFs under varied machining conditions for various materials should be investigated to identify any limitations or compatibility issues. Although BDFs are eco-friendly, it is important that they fulfil certification requirements and regulatory norms. In order to simplify regulatory compliance, future research should focus on verifying the safety and environmental credentials of BDFs.

2.8 CONCLUSION

The advancement of sustainability in EDM indicates a substantial move towards eco-friendly manufacturing approaches. The BDFs present a viable alternative to HBDFs due to their advantages over HBDFs, as discussed in an earlier section. Due to the biodegradability, non-toxic composition, and eco-friendly features of BDFs, their incorporation as dielectric fluids in EDM reflects proactive measures to enhance workplace conditions and minimize health hazards for personnel, along with improvements in environmental sustainability. In addition to the utilization of BDFs as dielectric fluids in EDM techniques, various other techniques, such as dry EDM and PMEDM, also contribute to sustainability through several innovations. The research findings indicate that the use of BDFs inEDM outperformed hydrocarbon oils and synthetic dielectric fluids, not only in machining performance but also in terms of social and environmental sustainability. This is because BDFs decompose readily and pose harm to personnel's health and working environment.

Regardless of the various advantages, various aspects need to be explored to make it socially, environmentally, and economically viable while utilizing BDFs in the EDM technique for large-scale commercial or industrial applications. Research and development efforts should focus on enhancing the durability of potential BDFs, improving machining effectiveness, and ensuring the overall sustainability of the EDM technique. To deal with these challenges, a thorough examination of the formulation of BDFs, various tests of their performance in real machining conditions, and the development of specialized equipment tailored according to the unique properties of BDFs are essential. Moreover, researchers, industry stakeholders, and policymakers should work together to optimize the benefits of BDFs, resolve regulatory concerns, and encourage wider use for enhanced efficiency and sustainability. Ongoing research on the utilization of BDFs may lead to the successful adoption of more sustainable EDM for large-scale commercial or industrial applications in the near future.

REFERENCES

[1] S. K. Anand and D. Chakradhar, "Hybrid and non-traditional manufacturing of bio-implants," in *Bioimplants Manufacturing*, edited by P. M. Abhilash, K. K. Gajrani, and X. Luo, Eds., Boca Raton: CRC Press, 2024, pp. 158–182. doi: 10.1201/9781003509943-8.

[2] N. K. Singh, L. Yadav, and S. Lal, "Experimental investigation for sustainable electric discharge machining with Pongamia and Jatropha as dielectric medium," *Advances in Materials and Processing Technologies*, vol. 8, no. 2, pp. 1447–1466, Apr. 2022, doi: 10.1080/2374068X.2020.1860499.

[3] Z. H. Shah and Q. A. Tahir, "Dielectric properties of vegetable oils," *Journal of Scientific Research*, vol. 3, no. 3, pp. 481–492, 2011, doi: 103329/jsr.v3i3.7049.

[4] S. Chandrasekar and G. Montanari, "Analysis of partial discharge characteristics of natural esters as dielectric fluid for electric power apparatus applications," *IEEE Transactions on Dielectrics and Electrical Insulation*, vol. 21, no. 3, pp. 1251–1259, June 2014, doi: 10.1109/TDEI.2014.6832272.
[5] S. K. Anand and D. Chakradhar, "Evaluation of the machining performance of Inconel X-750 in the EDM process using the CoCoSo method," in *Processing and Fabrication of Advanced Materials*, vol. 2, 2024, pp. 279–298. doi: 10.1007/978-981-97-5963-7_19.
[6] S. K. Anand and D. Chakradhar, "Performance and surface integrity in the electrical discharge machining of nickel-based superalloys: Analysis and assessment," *Journal of Materials Engineering and Performance*, Nov. 2025, doi: 10.1007/s11665-025-12512-8.
[7] S. P. Sivapirakasam, J. Mathew, and M. Surianarayanan, "Effect of process parameters on the breathing zone concentration of gaseous hydrocarbons—A study of an electrical discharge machining process," *Human and Ecological Risk Assessment: An International Journal*, vol. 17, no. 6, pp. 1247–1262, Nov. 2011, doi: 10.1080/10807039.2011.618390.
[8] N. M. Abbas, N. Yusoff, and R. Mahmod@Wahab, "Electrical discharge machining (EDM): practices in Malaysian industries and possible change towards green manufacturing," *Procedia Engineering*, vol. 41, pp. 1684–1688, 2012, doi: 10.1016/j.proeng.2012.07.368.
[9] F. Modica, V. Marrocco, G. Copani, and I. Fassi, "Sustainable micro-manufacturing of micro-components via micro electrical discharge machining," *Sustainability*, vol. 3, no. 12, pp. 2456–2469, Dec. 2011, doi: 10.3390/su3122456.
[10] G. Rotella, T. Lu, L. Settineri, O. W. Dillon, and I. S. Jawahir, "Dry and cryogenic machining: Comparison from the sustainability perspective," in *Sustainable Manufacturing*, edited by G. Seliger, Berlin, Heidelberg: Springer Berlin Heidelberg, 2012, pp. 95–100. doi: 10.1007/978-3-642-27290-5_14.
[11] I. S. Jawahir, P. C. Wanigarathne, and Wang X., "Product design and manufacturing processes for sustainability," *in Mechanical Engineers' Handbook*, edited by M. Kutz, ch. 12, pp. 414–443, 2006.
[12] E. Uhlmann, T.-M. Schimmelpfennig, I. Perfilov, J. Streckenbach, and L. Schweitzer, "Comparative analysis of dry-EDM and conventional EDM for the manufacturing of micro holes in Si3N4-TiN," *Procedia CIRP*, vol. 42, pp. 173–178, 2016, doi: 10.1016/j.procir.2016.02.214.
[13] L. Liqing and S. Yingjie, "Study of dry EDM with oxygen-mixed and cryogenic cooling approaches," *Procedia CIRP*, vol. 6, pp. 344–350, 2013, doi: 10.1016/j.procir.2013.03.055.
[14] J. P. Agrawal, N. Somani, and N. K. Gupta, "A systematic review on powder-mixed electrical discharge machining (PMEDM) technique for machining of difficult-to-machine materials," *Innovation and Emerging Technologies*, vol. 11, Jan. 2024, doi: 10.1142/S2737599424400024.
[15] R. Nadda, C. K. Nirala, P. K. Singh, D. Lee, R. Kumar, and T. Singh, "An overview of techniques for monitoring and compensating tool wear in micro-electrical discharge machining," *Heliyon*, vol. 10, no. 6, p. e26784, Mar. 2024, doi: 10.1016/j.heliyon.2024.e26784.
[16] J. B. Valaki, P. P. Rathod, and C. D. Sankhavara, "Investigations on technical feasibility of Jatropha curcas oil based bio dielectric fluid for sustainable electric discharge machining (EDM)," *Journal of Manufacturing Processes*, vol. 22, pp. 151–160, Apr. 2016, doi: 10.1016/j.jmapro.2016.03.004.
[17] J. B. Valaki and P. P. Rathod, "Investigating feasibility through performance analysis of green dielectrics for sustainable electric discharge machining," *Materials and Manufacturing Processes*, vol. 31, no. 4, pp. 541–549, Mar. 2016, doi: 10.1080/10426914.2015.1070430.
[18] J. B. Valaki, P. P. Rathod, B. C. Khatri, and J. R. Vaghela, "Investigations on palm oil based biodielectric fluid for sustainable electric discharge machining," in Proceedings

of the *International Conference on Advances in Materials and Manufacturing (ICAMM 2016)*, Bankok, Thailand, pp. 29–30, 2016.

[19] V. Srinivas Viswanth, R. Ramanujam, and G. Rajyalakshmi, "A review of research scope on sustainable and eco-friendly electrical discharge machining (E-EDM)," *Materials Today: Proceedings*, vol. 5, no. 5, pp. 12525–12533, 2018, doi: 10.1016/j.matpr.2018.02.234.

[20] J. B. Valaki, P. P. Rathod, and C. D. Sankhavara, "Investigations on technical feasibility of Jatropha curcas oil based bio dielectric fluid for sustainable electric discharge machining (EDM)," *Journal of Manufacturing Processes*, vol. 22, pp. 151–160, Apr. 2016, doi: 10.1016/j.jmapro.2016.03.004.

[21] J. B. Valaki and P. P. Rathod, "Assessment of operational feasibility of waste vegetable oil based bio-dielectric fluid for sustainable electric discharge machining (EDM)," *The International Journal of Advanced Manufacturing Technology*, vol. 87, no. 5–8, pp. 1509–1518, Nov. 2016, doi: 10.1007/s00170-015-7169-0.

[22] P. S. Ng, S. A. Kong, and S. H. Yeo, "Investigation of biodiesel dielectric in sustainable electrical discharge machining," *The International Journal of Advanced Manufacturing Technology*, vol. 90, no. 9–12, pp. 2549–2556, June 2017, doi: 10.1007/s00170-016-9572-6.

[23] A. Talebian-Kiakalaieh, N. A. S. Amin, and H. Mazaheri, "A review on novel processes of biodiesel production from waste cooking oil," *Appl Energy*, vol. 104, pp. 683–710, Apr. 2013, doi: 10.1016/j.apenergy.2012.11.061.

[24] P. Sadagopan and B. Mouliprasanth, "Investigation on the influence of different types of dielectrics in electrical discharge machining," *International Journal of Advanced Manufacturing Technology*, vol. 92, no. 1–4, pp. 277–291, Sep. 2017, doi: 10.1007/S00170-017-0039-1.

[25] S. Das, S. Paul, and B. Doloi, "An experimental and computational study on the feasibility of bio-dielectrics for sustainable electrical discharge machining," *J Manuf Process*, vol. 41, pp. 284–296, May 2019, doi: 10.1016/j.jmapro.2019.04.005.

[26] B. Singaravel, K. C. Shekar, G. G. Reddy, and S. D. Prasad, "Experimental investigation of vegetable oil as dielectric fluid in Electric discharge machining of Ti-6Al-4V," *Ain Shams Engineering Journal*, vol. 11, no. 1, pp. 143–147, Mar. 2020, doi: 10.1016/j.asej.2019.07.010.

[27] B. Singaravel, K. C. Shekar, G. G. Reddy, and S. D. Prasad, "Performance analysis of vegetable oil as dielectric fluid in electric discharge machining process of inconel 800," *Materials Science Forum*, vol. 978, pp. 77–83, 2020, doi: 10.4028/WWW.SCIENTIFIC.NET/MSF.978.77.

[28] B. P. Mishra and B. C. Routara, "Evaluation of technical feasibility and environmental impact of Calophyllum Inophyllum (Polanga) oil based bio-dielectric fluid for green EDM," *Measurement*, vol. 159, p. 107744, Jul. 2020, doi: 10.1016/j.measurement.2020.107744.

3 Unveiling the Evolution of Powder-Mixed Electrical Discharge Machining
Review and Future Directions

Deepti Ranjan Sahu, Manchu Mohan Krishna Sai, and Amitava Mandal

3.1 INTRODUCTION

The material removal ability of sparks occurring between two electrodes was noticed by Benjamin Priestley in 1751. Later, in the 1940s, Lazarenkos and Zolotykh made the systematic use of sparks for material removal. However, D.W. Rudorff is considered the first person to get the patent for Electrical Discharge Machining (EDM) in 1950 [1]. Because of the non-contact and thermo-electrical mechanism, spark erosion-based machining methods can machine electrically conductive materials of any hardness [2,3]. In EDM, the primary material removal mechanism for metals and alloys is melting and vaporization. However, for materials with higher melting and boiling points, e.g. ceramics, material spalling becomes the predominant material removal mechanism [4]. The EDM process has wide-ranging applications in the production of components for the surgical, automotive, nuclear, and aerospace industries. EDM is also employed abundantly in the mold, die, and cavity manufacturing industries due to its capability to create intricate and deep three-dimensional (3D) cavities [5–8]. Spark erosion machining encompasses three main methods: die-sink EDM, wire-cut EDM (WEDM), and micro EDM (μ-EDM). These methods typically employ hydrocarbon-based oils or deionized water as dielectric fluids. While hydrocarbon-based oils are prevalent in die-sink EDM, they pose significant environmental concerns. Moreover, EDM often involves a trade-off between achieving a high material removal rate (MRR) and preserving surface quality. Pursuing a high MRR may compromise surface quality, while prioritizing fine surface quality can reduce productivity. These challenges have led to the development of dry EDM and additive mixed EDM (AMEDM). Dry EDM helps mitigate environmental impact, whereas AMEDM holds promise for achieving both high MRR and fine surface quality[4]. The additives used in AMEDM can be of solid, liquid, or gaseous type. Gaseous type additives include nitrogen and oxygen, whereas liquid additives can be

DOI: 10.1201/9781003769675-3

surfactants, calcium- or urea-based solutions, etc. EDM with solid additives added to the dielectric in powder form is known as powder-mixed EDM (PMEDM).

3.1.1 Significant Applications of EDM

The basic EDM principle facilitates material removal via repetitive electrical discharges between the tool and workpiece. Appropriate modifications to this principle and the process environment have enabled the evolution of EDM into several advanced variants tailored for diverse industrial applications. The modified EDM process and the modified EDM setup are named based on the application. Some of these modifications of the basic EDM process are explained in this section.

(i) Drilling: A blind hole or through hole can be created on a workpiece by advancement of a rotating tool electrode of cylindrical shape toward the workpiece. This process is known as drilling EDM [9].
(ii) Milling: For machining a large surface, a revolving wheel with teeth taken as the tool is moved along the required path. This process is known as Milling EDM and can be used for complex 3D profiles [9].
(iii) Cutting: WEDM is an alteration available in the basic EDM that can be used for cutting operations. In WEDM, a wire (diameter: 0.03–0.3 mm) is fed to the machining zone concentric with a stream of dielectric fluid. It can be used for generating complex profiles on very hard materials. For cutting off or making a notch on the workpiece material of very high hardness, disk cutting EDM is used [10,11].
(iv) Die-Sink EDM: In die-sink EDM, a shaped tool is used for generating the desired profile on a workpiece of any hardness. The tool is made to have a shape complementary to that required on the workpiece. This same setup can be used for coating, also just by reversing the electrode polarity [9].
(v) Grinding: Grinding operation using electric discharges for material removal is known as electric discharge grinding [9].
(vi) Micro-EDM: Unlike conventional EDM, in μ-EDM, because of low energy discharges, material is removed in very small quantities. Hence, micro features can be generated on the workpiece with excellent accuracy [9].
(vii) Dry EDM: This involves the passage of high-pressure gas through a hollow cylindrical tool with a thin wall. This eliminates the production of any toxic fumes and also reduces any chance of fire hazards. Low MRR and process instability are some limitations of this process [12,13].

3.1.2 Limitations of Conventional EDM

(i) Electrical Discharge Machining (EDM) is limited to working with materials that conduct electricity. Non-conductive materials such as plastics or ceramics are not compatible with this method.
(ii) In contrast to conventional machining techniques such as milling or turning, EDM operates at a significantly slower pace. Material gets removed gradually via erosion, resulting in extended machining durations.

(iii) During the sparking process in EDM, the electrode is prone to deterioration, requiring frequent replacements or reshaping. This contributes to increased production expenses and downtime. Additionally, maintaining sharp corners and edges on the workpiece can pose challenges due to the erosion of the electrode.
(iv) The intense heat produced during sparking in EDM can result in the formation of a thin layer of resolidified material on the surface of the workpiece. This recast layer might exhibit properties different from the bulk material and might necessitate supplementary finishing processes.
(v) In conventional EDM, a dielectric fluid is used to insulate the space between the electrode and the workpiece. The disposal of this fluid can pose challenges due to the risk of contamination.

3.1.3 Introduction to the PMEDM Process

Erden and Bilgin are credited for pioneering the PMEDM process. In 1980, they conducted a study on the influence of various powder additives added to kerosene dielectric on MRR in EDM. Powder incorporated into the dielectric fluid leads to a reduction in dielectric breakdown strength, but it simultaneously boosts process stability, ultimately contributing to enhanced machining efficiency in PMEDM [14]. Since then, researchers have been progressively exploring the utilization of additive mixed dielectric fluids as an alternative to the conventional dielectric fluid. These additives can be solid, liquid, or gaseous. When solid additives are used, the process is termed as PMEDM. PMEDM offers the advantage of modifying the surface while carrying out the machining operation. The use of various powders, varying in concentration and size (ranging from micro to nano), has the potential to result in a higher removal rate and improved surface quality with minimum defects [4,15].

This chapter provides an introduction to EDM and PMEDM processes and discusses their working mechanisms, components of the EDM system, modeling aspects of EDM and PMEDM processes, current research trends, challenges, and scope for further research in improving the PMEDM process.

3.2 MECHANISM OF ELECTRICAL DISCHARGE MACHINING

EDM is a nonconventional thermoelectric machining procedure that removes material from conductive materials with any hardness [16,17]. By applying a potential difference across the tool and workpiece, intermittent and repetitive sparks are generated between them. The heat content of the sparks is utilized in the erosion of the electrode material by liquefying and vaporizing. The shape and dimension of the machined profile on the workpiece are restricted by that of the tool [11]. A schematic of a typical EDM setup is shown in Figure 3.1.

3.2.1 Discharge Initiation and Types of Discharges in EDM

Usually, from the inception of the EDM principle, it is believed that the role of negatively charged particles dominates the removal phenomena. When a voltage is

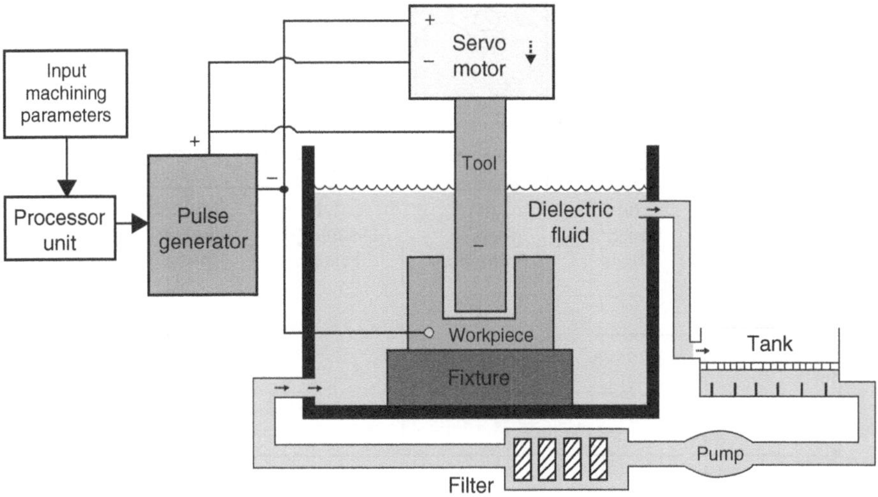

FIGURE 3.1 Schematic of a typical EDM setup [4,18].

applied across the electrodes, the free electrons on the cathode experience electrostatic forces. When this electrostatic force on the electrons exceeds their bonding energy with respect to (w.r.t.) the cathode, the free electrons leave the cathode and get accelerated toward the anode. During this accelerated movement, the electrons ionize the dielectric medium, generating ions and electrons. Gradually, with the generation of more number of electrons and positive ions, the spark gap region is concentrated with electrons and positive ions. Subsequently, the concentration of electrons and positive ions in the spark gap region becomes very high. This region, highly populated with electrons and positive ions, known as plasma, offers negligible electrical resistance, allowing movement of electrons from cathode to anode and positive ions from anode to cathode, resulting in an avalanche motion of electrons. This motion is visible as the spark. Thus, the electrical energy gets converted to kinetic energy (KE) of the charged particles, and when these charged particles collide with the electrodes, their KE gets converted to thermal energy, increasing the local temperature beyond 10,000°C. Because of this extreme temperature, material from the electrodes gets melted and vaporized. When the application of potential difference is stopped during the pulse interval, the spark puts off and the plasma channel collapses. On the collapse of the plasma channel, shock waves are generated, and a sudden inflow of dielectric fluid takes place toward the spark gap region. This flushes out the molten material generating the craters. This entire process has been illustrated in Figure 3.2a. Some molten material resolidifies, forming the recast layer [19]. Since the positively charged particles are nearly 10^4 times heavier than the negatively charged particles, they move with lesser velocity as compared to the negatively charged particles. Therefore, the KE being proportional to the square of velocity ($1/2*mv^2$) becomes higher for the negatively charged particles than that for the positively charged particles. Hence, more heat is generated at the anode due to the collision of the negatively charged particles coming from the cathode than the heat

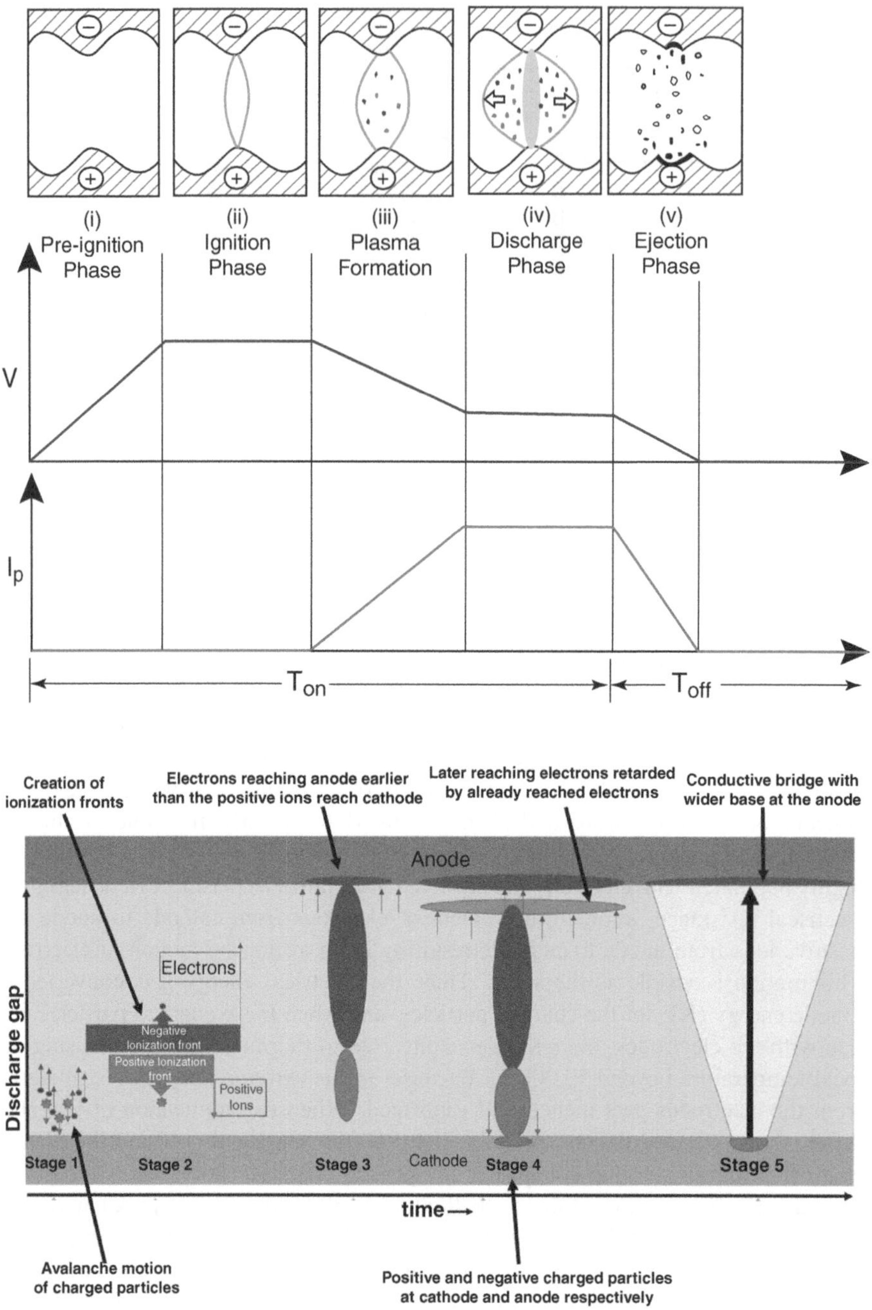

FIGURE 3.2 (a) Stages of material erosion in EDM [21] and (b) stages during discharge in EDM [22].

generated at the cathode by the collision of the positively charged particles coming from the anode and colliding with the cathode. This results in higher removal from the anode in the EDM process [11,20].

The mechanism discussed above is the traditionally accepted mechanism for EDM. However, another perspective for the material removal phenomena in the EDM process is illustrated in Figure 3.2b. The discharge process between the electrodes can be considered as a three-phase phenomenon. Phase one begins with the striking of electrons coming from the cathode with molecules of the dielectric and the production of more electrons and positively charged particles. More such collisions produce more number of electrons and positive ions. Once a threshold number of such collisions takes place, phase two is reached in which a positive ionization front and a negative ionization front are created just adjacent to the cathode and anode, respectively. The electrons being lighter hit the anode earlier than the positive ions hit the cathode and heat the surface. Now phase three is reached when the later coming electrons are retarded in front of the anode surface by the electrons that have previously hit the anode, and thus they are only capable of mildly heating the anode surface. Meanwhile, the positive ionization front also reaches the cathode. Thus, an electrically conductive bridge is formed between the cathode and anode. Since the electrons are lighter, they reach the anode earlier than the positive ions reach the cathode. When more number of electrons reach the anode, the earlier-reaching electrons are closer to the cathode. The later-coming electrons are retarded by the already-reached electrons and thus pushed sideways. Thus, the negative charge front is wider near the anode. Likewise, the positive ions reach the cathode. When the first batch of positive ions reaches the cathode, the conductive bridge is formed between the already existing negative ionization front near the anode and the positive ionization front just formed near the cathode. Even if more positive ions reach the cathode later on, they don't participate in the formation of the conductive bridge, because the conductive bridge is already closed with the first batch of positive ions near the cathode and the negative charge front near the anode. Because of this, the surface of this conductive bridge is wider on the side of the anode than on the side of the cathode. Thus, the current density at the anode becomes significantly lower than that at the cathode, resulting in a mild heating at the anode but intense heating at the cathode. Hence, the cathodic removal rate exceeds the anodic removal rate. Formation of more than one cathodic charge front is also possible on the cathode with the later-coming positive ions. But in this case of multiple conductive bridges, the conductive bridge with the least electrical resistance will sustain, and the others would collapse. With geometrically favorable positions, more than one conductive bridge can also join to form a stronger conductive bridge. Depending upon the pulse duration, the anodic removal may exceed the cathodic removal. In micro-EDM, the pulse duration is usually lower (less than 1 μs) and anodic workpiece gives a higher removal rate as compared to a cathodic workpiece. But with higher values of pulse durations in macro-EDM, cathodic removal becomes higher than anodic removal beyond a particular value of pulse duration. This particular value of pulse duration is again material dependent [22].

Types of discharges in the EDM process:

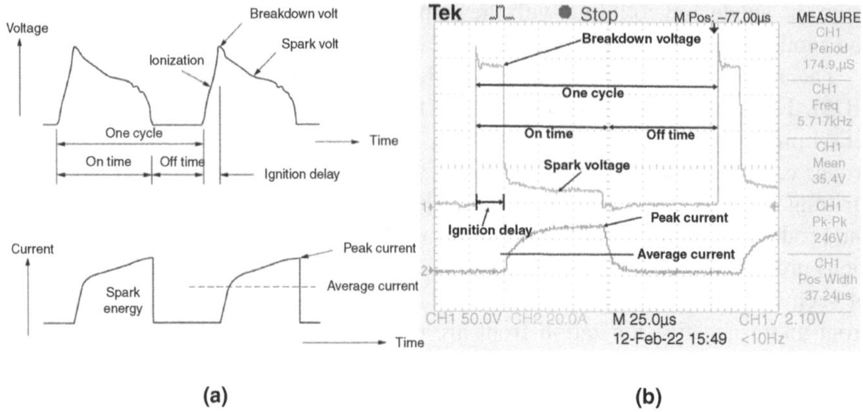

FIGURE 3.3 (a) Typical pulse during EDM [20] and (b) pulse during an actual EDM experiment.

A typical pulse during the EDM process is shown in Figure 3.3. Usually, five types of pulses are observed during EDM discharges. They are as follows:

a. Open circuit: This pulse is characterized by high gap voltage with a negligibly small gap current. This pulse is not capable of material removal.
b. Spark: This pulse consists of a portion with the discharge voltage along with some portion with high-frequency signals and a portion for dielectric breakdown. This is the most effective pulse type for material removal.
c. Transition arc: This pulse type is similar to the spark type of pulse but with a smaller component of high-frequency signals and no prominent component for dielectric breakdown.
d. Stable arc: This pulse type has no component of high-frequency signals and no component of dielectric breakdown. The discharge voltage is slightly lesser than the spark type of pulse, and the current is of similar magnitude as that of the spark and transition arc.
e. Short circuit: This pulse type is characterized by very small voltage and high current [23].

Uniformly distributed sparks on the tool and workpiece surface cause uniform material removal from the entire machining region, producing an even machined surface. Arcs occur at a localized region and remove material from that particular region only, thus resulting in an uneven machined surface. Arcs occur because of the incomplete deionization of the dielectric from the previous pulse. Spark is desired in EDM, but arc is not desired [19]. Figure 3.4 illustrates the types of pulses encountered during EDM.

3.2.2 EDM System and Its Components

An EDM setup consists of the following sub-systems:

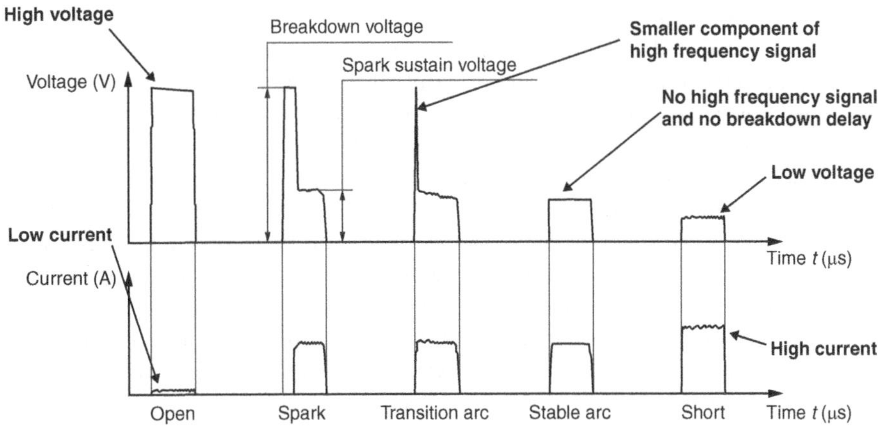

FIGURE 3.4 Types of pulses during the EDM process [23].

a) The power supply unit
b) Dielectric circulation system
c) The electrodes
d) The servo system
e) The work tank [11].

a) The power supply unit:
The supplied alternating current (AC) is transformed to pulsed direct current (DC) by the power supply unit and is utilized in generating sparks in the tool-workpiece gap. The power supply unit also senses the tool-workpiece gap voltage, which is used for controlling the servo system. The primary element within the power supply unit is the pulse generator. The pulse generators typically employed for EDM include:
   (i) Resistance-capacitance (RC)-type pulse generator
   RC generators can produce short pulses of constant discharge energy. Using high peak current with these short pulses, smaller craters can be generated for efficient finishing operation. Hence, these generators are mostly employed for micro-machining and finish machining operations. However, for RC generators, the time required for charging the capacitor is comparatively high. Hence, such pulse generators cannot produce high-frequency pulses because of which they can yield low MRR [24].
   (ii) Transistor-type pulse generator
   Transistor-type generators are capable of delivering high-frequency sparks. Because of the high spark frequency, high values of MRR can be obtained. Therefore, for rough machining processes, transistor-type pulse generators are most suitable. Usually, these generators are employed in conventional EDM setups [25].
b) Dielectric circulation system:
The dielectric circulation system comprises dielectric fluid, dielectric container, pump, delivery pipes and valves, and filter.

An adequately viscous fluid with good electrical insulating capability is often treated as an ideal dielectric fluid. The primary purpose of using a dielectric fluid is to provide an insulating medium between the tool and workpiece so that the sparks occur in a controlled manner. Second, the dielectric fluid carries away the heat generated by sparks and maintains the dielectric fluid temperature below its flash point. The third and important task of a dielectric fluid is flushing the debris generated by the erosion of the tool and workpiece, thereby avoiding the risk of short-circuiting. Flushing can be done by making a hole through the tool or workpiece and applying pressure or suction to the dielectric fluid. But the most commonly used flushing technique is jet flushing. Apart from hydrocarbon oils, another typically used dielectric fluid is deionized water. Deionized water is environment-friendly and produces a fine surface but causes high tool wear. Hence, for EDM, instead of deionized water, hydrocarbon oil is mostly preferred as a dielectric fluid, whereas WEDM employs deionized water as dielectric [26,27].

The pump takes the dielectric fluid from the reservoir through pipes and valves and delivers it to the machining zone. The dielectric clears the debris and passes through the work tank drain to the filter. The debris is retained by the filter, and clean dielectric goes back to the reservoir [11].

c) The electrodes:
Tool and workpiece are the electrodes in an EDM system between which the sparks are generated. Workpiece is the one on which the final cut profile is made. Tool is the one whose shape decides the shape of the cut profile generated. The material that promotes minimum tool erosion, easy handling and machinability, availability at a reasonable price and has good electrical conductivity is preferred to be used as a tool. Copper, copper tungsten, graphite, brass, and copper graphite are often used as tool materials for EDM. Copper and brass are easily machinable, so they provide the ease of making intricate tool shapes. But their limitation is a high wear rate. Copper tungsten, graphite, and copper graphite provide minimum tool wear. But copper tungsten is difficult to machine, whereas graphite tools are very brittle, making their handling on the shop floor a delicate job. So, depending upon the shape of the cut profile, allowed overcut of the cut profile, and production rate, the tool material is selected [11,20].

d) The servo system:
The servo system controls the tool's motion for maintaining appropriate separation between the tool and workpiece to match the MRR accurately. Signals from the gap voltage sensor control the operation of the servo system. Whenever any electrically conductive material gets trapped between the inter-electrode gap (IEG), the gap voltage sensor system detects this and responds by retracting the tool until the IEG is cleared. Once the IEG becomes clear, tool movement resumes and machining process continues. The functioning of the servo system greatly depends on the flushing system used. Ineffective flushing can keep the servo system busy in tool retraction, thereby making less time available for machining, whereas an effective

flushing technique can minimize the time spent in tool retraction and result in a fast machining process [11].

e) The work tank:
The work tank facilitates the housing of the workpiece. While using a hydrocarbon-based dielectric fluid, to avoid fire hazards, the sparking zone needs to be submerged inside the dielectric. The work tank is filled with dielectric to a level sufficient enough to submerge the tool-workpiece gap where the sparks occur. This level of dielectric is initially set by the operator, thereafter maintained by adjusting the valve openings for flushing inlet and drain.

3.2.3 Process Parameters of the EDM Process

a) Electrical parameters:
   (i) Peak current (I_p):
   I_p is the pre-set value of maximum current attainable during a discharge in the EDM process. But actually, the discharge current does not remain constant throughout the discharge; rather, it keeps on fluctuating. Hence, for most of the calculations, the average current or effective discharge current is used instead of peak current. Average current is the product of I_p and duty factor (D_f). With an increase in I_p, MRR and tool wear rate (TWR) increase, whereas surface finish deteriorates [28].
   (ii) Pulse on time (T_{on}):
   T_{on} is the time span for which current flows in one cycle during the EDM process [28].
   (iii) Pulse off time (T_{off}):
   T_{off} is the duration between two successive pulses. During T_{off}, the spark puts off, the tool retracts, and the dielectric fluid flushes the tool-workpiece gap, thus preventing short-circuiting and facilitating the uninterrupted functioning of the material removal process. [28].
   (iv) Duty factor (D_f):
   T_{on} and T_{off} together constitute a total cycle time. D_f is the proportion of cycle time taken up by T_{on} [28].

$$\text{Duty factor}(\%) = \left(\left(T_{on}\right) / \left(T_{on} + T_{off}\right)\right) * 100$$

   (v) Gap voltage (G_v):
   The voltage across the IEG is referred to as gap voltage for an EDM setup. Again, this gap voltage has different values at different stages of EDM discharge. In no spark condition, it is termed as open circuit voltage. At a particular value of this gap voltage, the dielectric starts to allow the passage of current. This value of gap voltage is known as breakdown voltage. After this, the gap voltage decreases and becomes stable at a certain value to sustain the spark. The value of gap voltage is known as discharge voltage or spark voltage.

(vi) Discharge energy:
Discharge energy in the EDM process primarily depends on the voltage, current, and duration of a discharge. By using a digital storage oscilloscope, the values of these parameters can be determined and used for determining the discharge energy. In general, greater discharge energy tends to result in an increased MRR.

(vii) Polarity:
Polarity in EDM refers to the terminals of the power supply to which the tool and workpiece are connected. The polarity is named after the workpiece, i.e. the polarity can be cathodic or anodic depending upon whether the workpiece is considered as cathode or anode, respectively. Another popular convention for naming the polarity is normal polarity or reverse polarity. In normal polarity, the tool and workpiece may be treated as anode and cathode, respectively, and in reverse polarity, the tool and workpiece can be cathode and anode, respectively. The convention for designating normal or reverse polarity can differ among manufacturers, with some choosing to designate either the tool or workpiece as the anode and the other as the cathode. The selection of polarity depends upon the pulse duration used and the material of the electrodes. For pulse duration in the order of microseconds, usually the tool is made anode and the workpiece is made cathode. Again, whether the electrode materials are pure metals or alloys also affects which electrode should be made as anode or cathode. The overall selection of polarity is decided in such a way that the material removal should be maximum for the workpiece and minimum for the tool [22].

b) Non-electrical parameters:
Electrode lift time, flushing velocity, tool shape and size, tool material, and dielectric type are the non-electrical parameters that affect the overall performance of the EDM process [4].

3.2.4 Performance Measures of EDM Process

(i) Material removal rate (MRR)
The productivity of the EDM process is measured by MRR. Hence, parameters of the EDM process are often optimized to yield high MRR when productivity is the prime objective. MRR is defined as material removed in unit time, in terms of either mass or volume.

(ii) Tool wear rate (TWR)
TWR is a quantification of the material erosion rate from tool. TWR becomes a crucial performance parameter when the dimensional accuracy of the machined profile is of prime importance for a specific application. Typically, TWR is expressed in mass removal per unit time or volume removal per unit time.

(iii) Surface roughness
Surface roughness (SR) refers to the presence of closely positioned irregularities on the machined surface. SR mainly depends on the morphology of

craters formed and the nature of their distribution, which again is a function of discharge energy [29].

(iv) Surface integrity
Surface integrity is a performance indicator utilized for evaluating the quality of the machined surface. It comprises sub-surface quality parameters, viz. the extent of heat-affected zone, recast layer thickness, density and depth of cracks on the machined surface, changes in microstructure, composition, hardness. In the EDM process, heat-affected zone refers to that portion of the workpiece which remains un-melted but undergoes a phase transformation resembling the effects of heat treatment processes due to exposure to the elevated temperatures generated by electrical discharges. The unflushed resolidified layer above the heat-affected portion on the workpiece is termed as recast layer. Residual stresses develop on the machined surface due to rapid heating and subsequent quenching; when these stresses exceed the tensile strength of the material, surface cracks are formed. For quantifying these cracks, the depth and density of these cracks are used [17,30].

3.3 RESEARCH PERSPECTIVES IN ANALYSIS OF THE MECHANISM OF MATERIAL REMOVAL IN EDM

The fundamental mechanism of EDM and its variants involves a multi-physics phenomenon occurring over a very small duration of time. This complex phenomenon is difficult to interpret from mere observation of macroscopic quantities, viz. MRR or TWR. In this context, modeling of the process can act as a tool for accurate interpretation of the underlying process mechanism. Usually, two types of modeling approaches are followed by researchers in modeling spark erosion-based machining processes. They are:

(i) Data-based modeling is mostly targeted toward the representation of the process in terms of regression equations. These models usually ignore the underlying physics. The regression equations and predicted values are based on an available set of data. They lack accuracy beyond the range of the data considered for developing these models [31].
(ii) Physics-based models are mostly developed on the basis of the principles of physics governing that particular process. Since they are developed on the basis of physics rather than an available data set, the reliability of their prediction is independent of the range of data and remains valid till the governing physics remains valid in that data range.

The existing physics-based models for spark erosion–based machining processes are generally categorized into three main approaches: finite element method based models, molecular dynamics models, and flow-field-based models. Molecular dynamics simulations help in the evaluation and prediction of statics as well as dynamic behavior at an atomic or molecular level for a material undergoing the changes during its processing. Formation of stacking faults, dislocations, etc.

in a material processed by the spark erosion process can be simulated with the help of molecular dynamics simulation [32–34]. Flow-field simulation uses the principle of computational fluid dynamics in simulating the flow of dielectric fluid, bubbles, and debris in the tool-workpiece gap during the EDM process [35–37]. FEM-based models can help in predicting the temperature distribution in materials processed by EDM, which can further be used to predict other responses [38–40]. As shown in Figure 3.5, the majority of the physics-based models reported in the available literature are FEM-based. Hence, the discussion in this section is mainly focused on FEM-based models. Steps followed in a typical FEM-based model for spark erosion-based machining processes are illustrated in Figure 3.6. First, in a generalized FEM model used for multi-spark simulations, it is crucial to ascertain the discharge location. Second, irrespective of whether it is a single-spark or multi-spark model, the vital model conditions, viz. heat input, spark radius, energy distribution ratio (EDR), and flushing efficiency, are defined. Lastly, by employing this generalized FEM model, the temperature distribution is estimated, which in turn allows for the prediction of various machining performance measures. The transient 3D heat conduction equation without heat generation governing the FEM-based simulation of spark erosion-based machining processes is given by Equation 3.1 [41–43].

$$\frac{1}{r}\frac{\partial}{\partial r}\left(K(T)r\frac{\partial T}{\partial r}\right)+\frac{\partial}{\partial z}\left(K(T)\frac{\partial T}{\partial z}\right)=\rho c_p\frac{\partial T}{\partial t} \tag{3.1}$$

Here, r represents radial distance and z represents vertical distance in the cylindrical coordinate system. $K(T)$, ρ, and c_p are the thermal conductivity, density, and specific heat of the workpiece material. $K(T)$ is a function of temperature. 't' indicates time and 'T' is the temperature field variable.

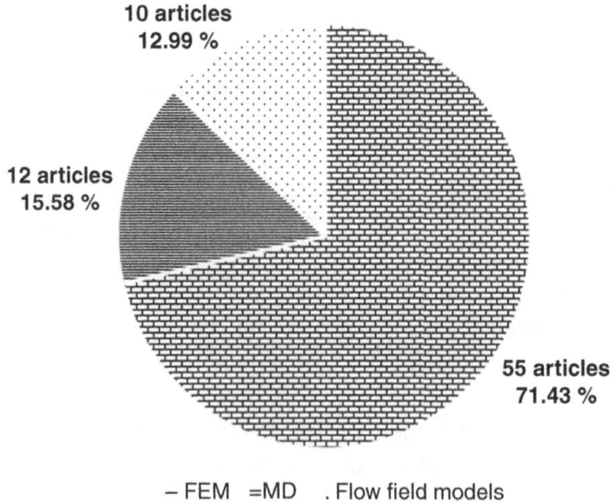

FIGURE 3.5 Pie chart representing the percentage of articles on different modeling approaches for EDM till 2022 [44].

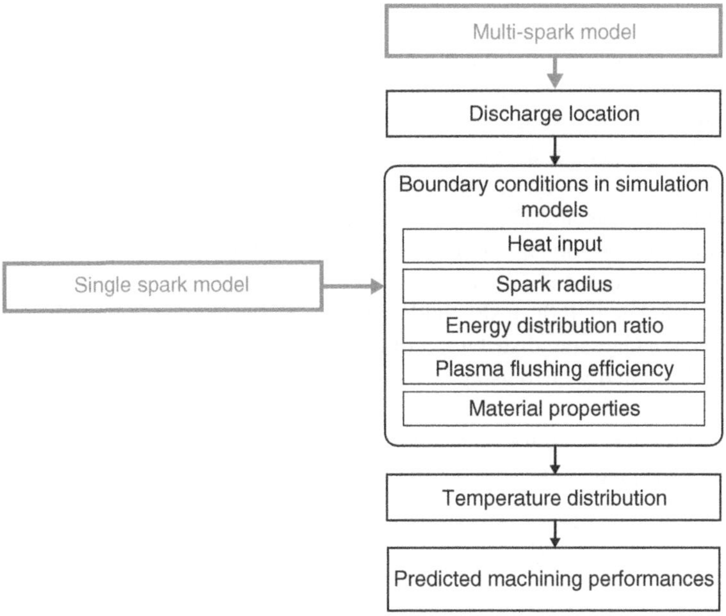

FIGURE 3.6 Steps followed in a typical FEM-based model for spark erosion-based machining processes [44].

3.3.1 Approach for Locating the Discharges

In the case of single-spark simulation, locating the discharge is not a concern [41,45,46]. But for multi-spark simulations, there is randomness in almost all aspects of the discharge phenomena. The locations where the sequence of multiple discharges occurs are not deterministic but occurs in a random manner. Defining a mechanism for locating the discharges is crucial. The energy content of each discharge is not definitive, rather varies in a random manner. Moreover, at a particular instant, the total number of discharges is not fixed and can vary randomly. Incorporating all these random patterns in a model is a very complex task [38,47–49]. Until the tool remains flat, the position of the next discharge can be anticipated to occur at the point where the electric field on the workpiece surface is at its maximum [50]. The formed bubbles and debris during discharges in EDM can affect the discharge locations [51,52]. A multi-spark EDM model with all these factors taken care of can predict MRR and SR with error as low as 3% and 6%, respectively[53]. Thus, multi-spark simulation can be more accurate but is computationally expensive and reduces the efficiency of simulation.

3.3.2 Boundary Conditions

Irrespective of a model being single-spark or multi-spark, the following factors constituting the boundary conditions are of utmost importance for the prediction accuracy of an FEM-based thermo-physical model of the EDM process.

3.3.2.1 Heat Source

The major characteristic features of the heat source used for modeling of EDM are shape, distribution of heat intensity, i.e. heat flux, and heat source radius. The commonly used heat source types for modeling the EDM process are shown in Figure 3.7. Dibitonto et al. [54] used a point heat source in modeling the cathode erosion process. Dijck and Dutre [45] used a time-dependent uniform heat flux in their model. Snoeys and Dijck [55], Beck [56,57], and Tariq et al. [58] used a disk heat source in their model. Pandit and Rajurkar [40] used a circular heat source for their model. Pandey and Jilani [59] used a disc-shaped heat source in their model. However, Patel et al. [60], Joshi and Pande [61], Bergs et al. [62], and Shabgard et al. [63] adopted a Gaussian distribution of heat intensity with maximum intensity at the center and gradually decreasing toward the periphery in their model and obtained better accuracy.

The expressions used for representing the Gaussian heat flux in the EDM process are given by Equations 3.2 and 3.3 [64].

$$q(r) = q_0 \times \exp\left(-4.5\times\left(\left(r / r_p\right)^2\right)\right) \tag{3.2}$$

$$q_0 = \left(4.45\times F_c \times V_d \times I_d\right) / \left(\pi \times \left(r_p\right)^2\right) \tag{3.3}$$

where *q(r)* describes the variation of heat flux w.r.t. the radial distance from the center of the plasma, r denotes the radial distance, q_0 is the maximum intensity of heat flux at the center of the discharge channel, r_p denotes the spark radius, F_c denotes the fraction of discharge energy shared by the cathode in case of cathode erosion model; in case of anode erosion model, F_c is substituted by the fraction of discharge energy distributed to the anode, i.e. F_a; V_d and I_d represent the discharge voltage and current, respectively. The exponent −4.5 describes the degree of flatness of the Gaussian heat

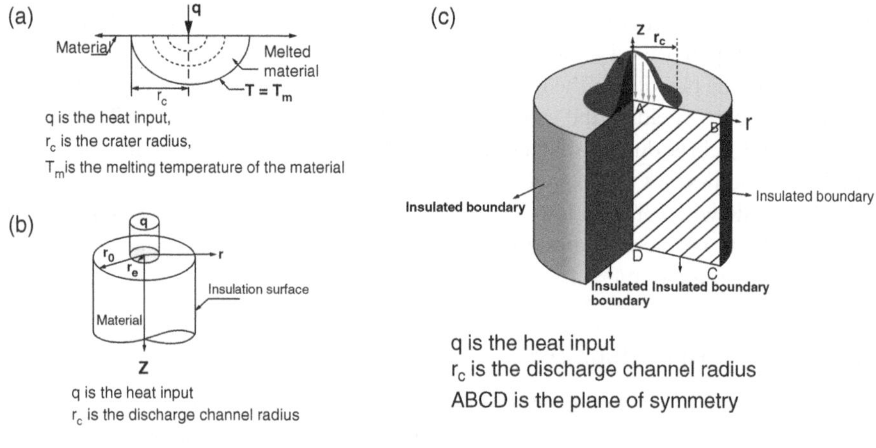

FIGURE 3.7 (a) Point heat source, (b) disk heat source, and (c) Gaussian heat source [44].

flux. As this exponent value increases (e.g. 0), the Gaussian heat flux distribution becomes flatter, whereas with a decrease in the value of this exponent, the Gaussian heat flux becomes steeper [65].

The spark radius (r_p) used by researchers in describing the heat source for modeling the EDM process was mainly of a constant type or an expanding type. Since the sparks are very short-lived in the EDM process, measurement of their radius is not easy. However, various researchers have used different equations for representing the r_p in modeling the EDM process. Erden considered r_p as a function of time and discharge power and used Equation 3.4 for spark radius in his model [66]. Patel et al. used spark radius as a function of pulse duration given by equation 3.5 in modeling the anode erosion process in EDM [60]. Zhang et al. also used Equation 3.5 for spark radius in modeling the EDM process [67]. Tan and Yeo used spark radius as a function of pulse duration but used a different expression as indicated by Equation 3.6 [68]. Shabgard et al. used expressions of spark radius as a function of pulse duration, but the expressions were different for cathode (Equation 3.8) and anode (Equation 3.7) [69]. Salonitis et al. considered spark radius as a function of discharge current and pulse duration, indicated by Equation 3.9 [70].

$$r_p = K' \times (E_{\text{dis}})^a \times (T_{\text{dis}})^b \tag{3.4}$$

$$r_p = 0.788 \times (T_{\text{dis}})^{0.75} \tag{3.5}$$

$$r_p = 0.059 \times (T_{\text{dis}})^{0.79} \tag{3.6}$$

$$r_p = 0.0284 \times (T_{\text{dis}})^{0.9115} \tag{3.7}$$

$$r_p = 0.0425 \times (T_{\text{dis}})^{0.0895} \tag{3.8}$$

$$r_p = 2.04 \times (I_{\text{dis}})^{0.43} \times (T_{\text{dis}})^{0.44} \tag{3.9}$$

where E_{dis} is the discharge energy, I_{dis} is the discharge current, T_{dis} is the discharge duration, and K', a, and b are empirical constants.

Among all these expressions for spark radius, Equation 3.9 used by Salonitis has been used by most of the researchers and has shown more accurate predictions [42,71–73].

3.3.2.2 Energy Distribution Ratio

In spark erosion-based processes, the discharge energy available at the machining zone is not fully utilized in material removal. Instead, only a fraction of this discharge energy gets utilized in material removal. The discharge energy actually

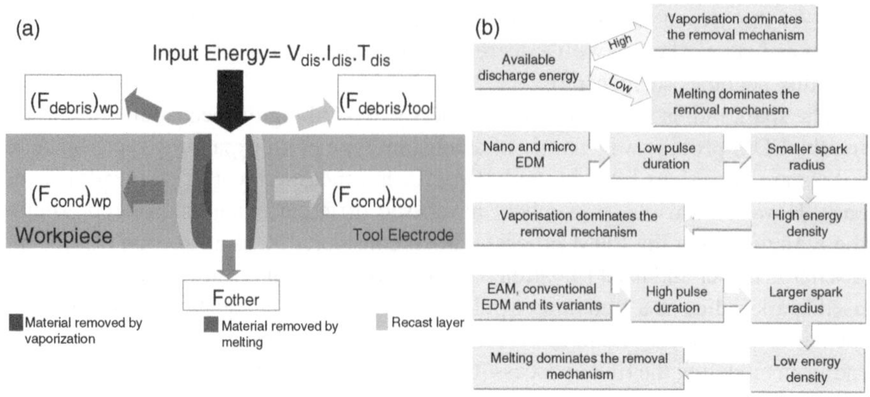

FIGURE 3.8 (a) Scheme of discharge energy distribution and (b) mechanism of material removal [75].

gets distributed into three components. One component gets distributed to the tool electrode. The second component gets distributed to the workpiece. The third component goes unutilized in the form of loss to dielectric fluid, loss in the form of sound and light of the spark. The component getting distributed to the electrodes has two components, i.e. the one getting wasted in conductive heat loss and the other getting utilized in material removal [74]. Again, the component of discharge energy that gets utilized in material removal can be due to melting or due to vaporization, depending upon the available discharge energy. Since the energy of vaporization is more than the energy of melting, if the available discharge energy is high, material erosion by vaporization dominates over material erosion by melting. Though melting and vaporization are the mechanisms of material removal in spark erosion-based machining processes, the dominating removal mechanism in micro-EDM is different than that in conventional EDM. In the case of micro-EDM, the spark radius is small because of the low value of pulse duration. Hence, due to high energy density, vaporization dominates the removal process. In conventional EDM, due to high pulse duration, the spark radius is comparatively larger, leading to lower energy density, and hence, in conventional EDM, melting dominates the material erosion phenomena. This scheme of discharge energy distribution is illustrated in Figure 3.8a [75]. In conventional EDM, a significant portion of the discharge energy is observed to be dissipated into the gap and is lost as heat conducted into the electrodes [76]. Zahiruddin and Kuneida's study [74] revealed that in the case of micro-EDM, a larger fraction of energy was effectively utilized for material removal from the workpiece compared to macro-EDM. Conversely, for macro-EDM, a higher proportion of energy was lost through conduction in the workpiece as compared to micro-EDM. Shen et al. [77] in their research discovered that the anode received a higher amount of discharge energy compared to the cathode. Xia et al. [78] in a single-spark study found that the energy fraction for the anode was slightly greater than that for the cathode. Hayakawa et al. [79] observed the power distributed to the anode to be greater than that to the cathode. A study by Singh [80] reported a variation in the energy fraction distributed to the workpiece (anode), ranging from 6.1% to 26.82%. DiBitonto et al. [81] reported that 18.3% of the discharge energy was distributed to the cathode and 8% to the anode. Yeo et al. [82] suggested that in their study on micro-EDM, 39% of the discharge energy was transferred to the cathode, while 14% went to the anode. Xia et al. [83] reported that the fraction of discharge energy transferred to the cathode and anode was 34% and 48%, respectively. Zahiruddin and Kunieda [84] investigated the energy distribution scheme in micro-EDM and found that the total discharge energy distributed to the anode varied from 6.6% to 10.8%, while that to the cathode varied from 3.3% to 3.7%. More than 85% of the discharge energy got wasted in the discharge gap. Xia et al. [85] in a single-spark EDM experiment reported that irrespective of polarity, the fraction of energy distributed to the anode and cathode was nearly equal and ranged from 15% to 30%.

Several parametric studies on the distribution of discharge energy in EDM have been reported. Shen et al. reported energy distribution in EDM as a function of T_{on}, inter-electrode separation, and shape of the electrode [77]. Zhang et al. also reported variation in the energy distributed to the tool and workpiece with respect to T_{on} [76]. The study by Singh [80] reported that the variation of fraction of energy distribution to the electrodes and dielectric with respect to T_{on} and current was not following a specific pattern; rather it was different at different combinations of T_{on} and current. But the energy distributed to the dielectric was always increasing with an increase in T_{on}. Zahiruddin and Kunieda [84] observed that with an increase in T_{on}, the total energy distributed to the electrodes, i.e. sum of energy distributed to the anode and cathode, was found to be increasing. Xia et al. [85] in a single-spark EDM experiment reported that the energy distributed to electrodes was independent of T_{on} but dependent on the IEG. Okada et al. [86], while experimenting on two dielectric mediums, i.e. deionized water and kerosene, reported that the influence of the dielectric medium on the energy distribution was considerable. They reported the conductive loss in electrodes and energy taken away by debris as a function of T_{on}. The fractions of energy distributed to the electrodes were found to be independent of T_{on} but dependent on current. Hayakawa et al. [79] observed that the power distributed to the anode was increasing with T_{on}. Energy lost to the dielectric also increased at a higher T_{on} value.

3.3.2.3 Plasma Flushing Efficiency

During the EDM process, some material gets removed from both of the electrodes and some material resolidifies and gets deposited on the electrodes as a recast layer. Plasma flushing efficiency (η_{PFE}) can be defined as the fraction of actual material removed from the electrodes during EDM to the theoretical material that should have been removed from the electrodes [87]. In other words, η_{PFE} in the EDM process is an indicator of the efficiency of plasma channel in removing the melted material from the tool electrode as well as the workpiece electrode [63]. Patel et al. [88] and DiBitonto et al. [81] initially used this term as a ratio of effectively removed material to the melting volume. Prior to DiBitonto et al., the same term was used as the metal removal efficiency by Van Dijck [89]. Shabgard et al. [63] have used η_{PFE} as a correction term to the crater dimensions obtained after the simulation to predict the final dimensions of the crater. Jithin et al. [87] obtained a more accurate prediction of the crater dimensions by using η_{PFE} as a correction term during the simulation itself.

3.3.2.4 Prediction from Model

In spark erosion-based machining processes, the material removal and other allied phenomena take place through thermal loading and subsequent cooling. Hence, for the majority of models developed for EDM, the primary purpose is to obtain the temperature distribution. From the temperature distribution, removal rate, surface texture, recast layer thickness, residual stress, etc. are estimated. Reviewing the available literature, it is observed that multi-spark simulations are complex and can adversely affect the computational efficiency, whereas single-spark simulations can deliver high computational speed by simplifying the model. Thus, though single-spark models face challenges in accurately simulating the 3D machined surface or modeling the macro material removal, they are easy to formulate and can predict the temperature distribution and some vital parameters.

3.4 UNVEILING THE ASPECTS OF POWDER-MIXED ELECTRICAL DISCHARGE MACHINING PROCESS

3.4.1 Overview of Powder-Mixed EDM Process

Erden and Bilgin are credited for pioneering the PMEDM process. In 1980, they conducted a study on the influence of various powder additives added to kerosene dielectric on MRR in EDM. Powder incorporated into the dielectric fluid leads to a reduction in dielectric breakdown strength, but it simultaneously boosts process stability, ultimately contributing to enhanced machining efficiency in PMEDM [14]. Since then, researchers have been progressively exploring the utilization of additive mixed dielectric fluids as an alternative to the conventional dielectric fluid. These additives can be solid, liquid, or gaseous. When solid additives are used, the process is termed as PMEDM process. PMEDM offers the advantage of modifying the surface while carrying out the machining operation. The use of various powders, varying in concentration and size (ranging from micro to nano), has the potential to result in a higher removal rate and improved surface quality with minimum defects [4,15].

3.4.2 Comparison of the Working Principles of EDM and PMEDM

For both conventional EDM and PMEDM, melting and vaporization caused by pulsating discharges are the fundamental working principles of material removal. Usually, in conventional EDM, on the application of a potential difference across the electrodes, the discharges initiate between the electrodes. But in PMEDM, due to the powders in the dielectric, the discharges take place in a different pattern. On the application of a potential difference, if the added powders are electrically non-conductive, then charges accumulate around the particles, and if the added powders are electrically conductive, then the charges accumulate within the particles. Accumulation of positive charges occurs closer to the cathode and that of

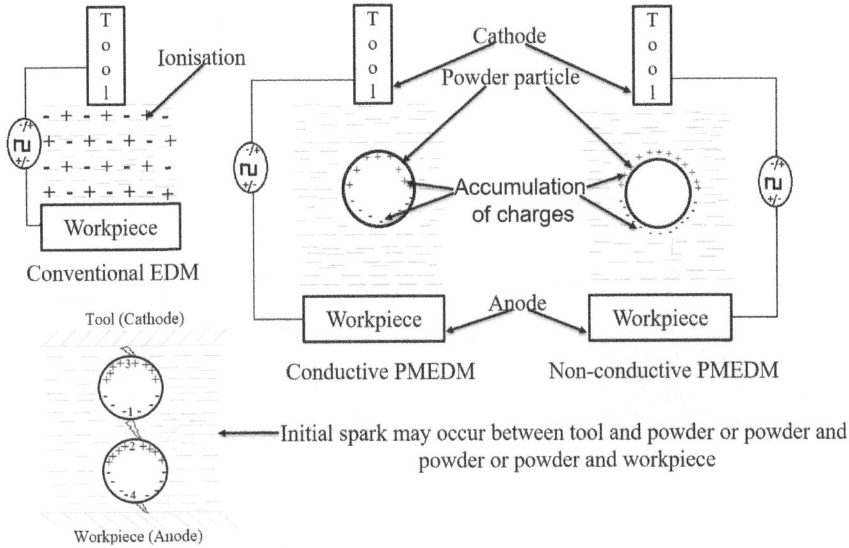

FIGURE 3.9 Schematic explaining the mechanism of conventional EDM and PMEDM processes [16,90,91].

negative charges occurs closer to the anode. Subsequently, the initial discharge may take place between the tool and a powder particle, between two particles of powder, or between a powder particle and the work material based on where the smallest spark gap is available. After the initial discharge, redistribution of discharges takes place, and the subsequent discharges occur in a similar manner. Thus, in PMEDM, the number of sparks increases as compared to the conventional EDM. So for a particular value of available discharge energy, discharge energy per spark in PMEDM is lesser than that in conventional EDM. This leads to the formation of smaller craters and a better surface finish in PMEDM as compared to conventional EDM. Again, since the number of sparks increases in PMEDM, MRR in PMEDM exceeds MRR in EDM. This comparative description of the working mechanism for conventional EDM and PMEDM (for reverse polarity with workpiece and tool as anode and cathode, respectively) has been illustrated through the schematic presented in Figure 3.9 [16,90,91].

As shown in Figure 3.10b, due to the elongated spark gap in PMEDM, the energy density is low as compared to that in conventional EDM (Figure 3.10a). Hence, the crater formed is deep with small diameter in the case of conventional EDM (Figure 3.10c) and shallow with larger diameter in the case of PMEDM (Figure 3.10d).

3.4.3 Process and Performance Parameters of PMEDM

Apart from the parameters of conventional EDM, size and concentration of the powder and type or properties of the powder, i.e. electrically conductive or non-conductive, play an important role in deciding the performance of the PMEDM process. The process and performance parameters of PMEDM have been illustrated in Figure 3.11.

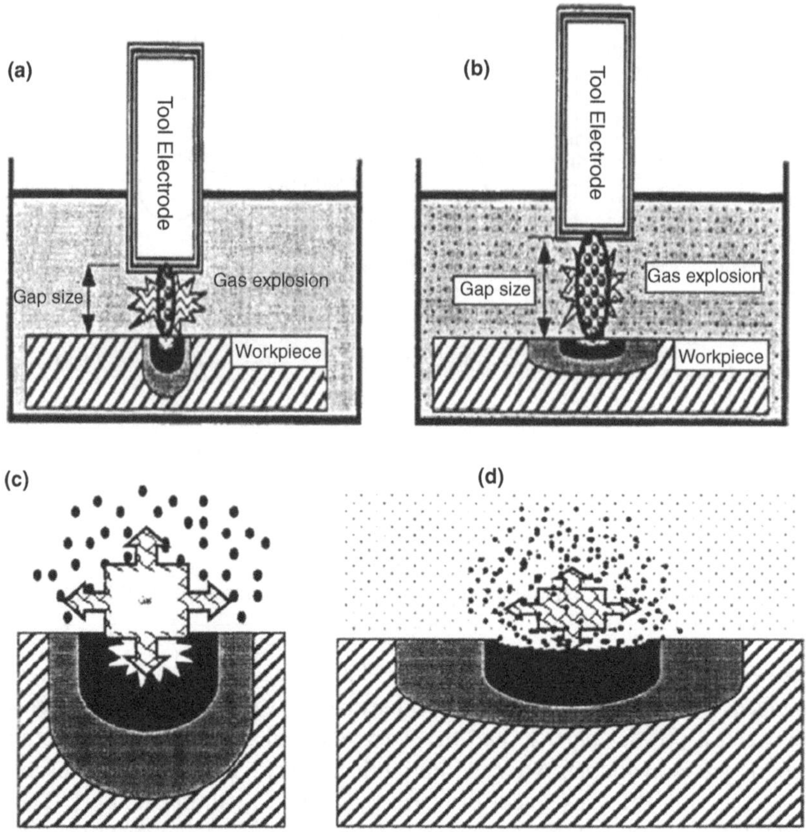

FIGURE 3.10 Material removal in conventional EDM and PMEDM [4,92].

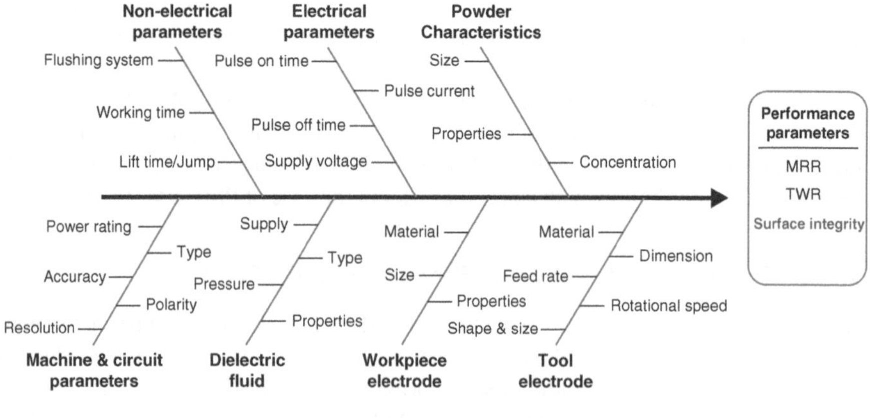

FIGURE 3.11 Cause and effect diagram for the PMEDM process [4,93,94].

3.4.4 Advantages of PMEDM

The following features of PMEDM make it advantageous over EDM.

(i) It can produce a mirror-like surface finish
(ii) The resulting recast layer in PMEDM is of uniform thickness and can have properties like corrosion and abrasion resistance
(iii) Reduced dielectric breakdown voltage and smaller ignition delay time
(iv) Better process stability
(v) Higher MRR
(vi) Enlarged and widened spark gap leading to less intense sparks and hence less surface cracks
(vii) Improved flushing efficiency due to enlarged spark gap [4,28,95,96].

3.4.5 Limitations of PMEDM

Although PMEDM presents notable benefits compared to conventional EDM, it does come with certain drawbacks. Some of the major drawbacks are:

(i) The presence of powder in the dielectric increases the process complexity in terms of the selection of powder type, concentration, size, making the optimization of process parameters more complicated.
(ii) Disposal of powder-mixed dielectric fluid is more challenging than that of conventional dielectric fluid.
(iii) Due to a lack of reusability, the powder additives used in PMEDM increase the overall machining cost.
(iv) There is uncertainty about how the powder additives would react with the surface of the material being processed. Depending upon whether the surface properties get improved or deteriorated, there may be a need for additional post-processing steps.
(v) As compared to the conventional EDM process, PMEDM is a recent process. Thus, the understanding of the complex interaction between the powder, discharge process, and workpiece material is very limited.

3.4.6 Applications of PMEDM

Machining and surface modification are the two major applications of the PMEDM process. Initially, the use of powder-mixed dielectric has gained popularity because of its ability to increase MRR and surface finish. Later on, depending upon the properties and composition of the added powder to the dielectric, specific properties, viz. corrosion resistance, abrasion resistance, lubricating ability, and higher hardness, could be imparted to the surface processed by PMEDM. Therefore, PMEDM has garnered widespread recognition and adoption in two key application areas, i.e. machining and surface modification. Machining of holes with large aspect ratio in exotic materials often used for the automobile industries, manufacturing of tools, molds, and dies, machining of titanium- and nickel-based alloys used in aerospace applications, and manufacturing of surgical equipment with complex shapes, bio-implants, and texture generation on implants. are some of the significant application areas where EDM and PMEDM processes are useful [28].

3.5 STATE OF THE ART IN PMEDM

Back in 1980, Erden and Bilgin attempted a study on the effects of incorporating powder into the dielectric for machining mild steel. They observed a rise in the machining rate as the concentration of impurities increased. Nonetheless, as the concentration of abrasive powder increased, it resulted in early breakdown, causing short-circuiting and a decline in machining performance [97]. Graphite micro powder of 10 μm size added to kerosene was seen to increase MRR and reduce TWR [1]. In a study using aluminum (Al), silicon (Si), and graphite (Gr) powder with concentrations ranging from 2 to 40 g/L, it was observed that as the concentration increased, the gap distance between electrodes also increased. However, there was no direct correlation between gap distance and surface finish. Notably, lower powder concentrations led to improved surface finishes when using Gr and Si powders [98]. A study conducted by Kobayashi et al. on PMEDM using Al, Si, and Gr powder revealed that Gr and Al powder could produce a finer surface as compared to Si powder [99]. But Mohri et al. claimed a mirror-finish surface-generating capability of Si powder[100]. Yu et al. reported the achievement of enlarged discharge gap and increased MRR in the machining of tungsten carbide by aluminum PMEDM [101]. Ming and He claimed a reduction in surface cracks and an increase in surface hardness by the use of conductive powder and lipophilic surface agents in PMEDM of high carbon steel [102]. Wong et al. reported the achievement of improved surface quality by incorporating Si powder into the working fluid in the machining of SKH-54 tool steel by EDM [103]. Uno et al. claimed improvement in abrasion resistance of the surface of aluminum bronze parts by incorporating nickel powder into the working medium in EDM [104]. Chow et al. machined a titanium alloy using Al and silicon carbide (SiC) powder-mixed EDM. High MRR, increased discharge gap, and better flushing were the key observations in their study [105]. Furutani was able to generate a very hard surface with hardness of the order of 1,600 HV on a carbon steel workpiece while machining it by titanium (Ti) PMEDM [106]. Peças et al. observed betterment in the surface quality by incorporating Si powder into the working medium at a concentration of 2 g/L while machining AISI H13 tool steel by the PMEDM process [107]. Klocke et al. witnessed the enlargement of plasma channel due to the addition of Al powder to the working medium. They claimed that the machined surface composition and morphology were dependent on the physical properties of the powder additives [108]. Using surfactant along with the added powder could bring about improvement in the issue of agglomeration and settling down of the powder in the Al PMEDM process. Apart from avoiding the agglomeration issue, betterment in surface finish was also observed [109]. Kansal et al., using response surface method (RSM)-based optimization of process parameters such as pulse duration, powder concentration, peak current, and duty factor in Si PMEDM, achieved improved MRR and surface [93]. In another RSM-based parametric study on PMEDM, Kansal et al. observed that the enhancement in MRR and TWR was dependent on the fineness of the added powder [110]. Yeo et al. carried out experiments encompassing a low range of discharge energies (2.5–25 μJ). They examined the effects of powder addition and observed notable differences in crater shapes. The introduction of powder resulted in the creation of smaller, circular craters [111]. Chow et al., in their 2008 study on micro-slit generation, explored the effectiveness of SiC powder-mixed water as the working fluid in EDM of Ti6Al4V alloy. Their findings indicated that the inclusion of SiC powder in pure water resulted in higher tool erosion and the production of larger slits. Moreover, using SiC PMEDM led to reduced burr formation [112]. Wu et al. conducted a study to assess the impact of surfactants in PMEDM of mold steel. By decreasing the tendency of powder agglomeration, the surfactants in the PMEDM process led to a 40% enhancement in MRR. But it also resulted in a deterioration of SR [113]. Kumar and Singh reported surface alloying capabilities of manganese (Mn) powder-suspended dielectric delivering higher hardness of the machined surface. The surface alloying phenomenon was favored by smaller values of current and pulse duration [114]. Ojha et al. conducted a study to evaluate MRR and TWR while machining EN8-carbon steel material with a dielectric containing chromium (Cr) powder. Their findings showed that higher powder concentration resulted in increased MRR and a larger tool diameter resulted in reduced tool wear [115]. In a study with Al powder-mixed distilled water as a dielectric, it was observed that positive polarity and negative polarity are conducive for higher MRR and better surface finish, respectively [116]. In a study on Gr, tungsten (W), and Si PMEDM, Bhattacharya et al. found that tungsten PMEDM with W-Cu tool could produce a machined surface with high microhardness whereas tungsten PMEDM with brass tool could result in better surface finish [117]. In a study on Al PMEDM, Goyal and Singh reported that extreme concentration of powder, i.e. very high or very low concentration of powder, can reduce MRR, whereas an increase in powder concentration resulted in improved surface finish [118]. In a parametric study on carbon nanotube (CNT)-mixed EDM, Kumar observed that a suitable selection of the concentration of CNT could result in the improvement of MRR as well as surface finish. In their study, concentration of CNT and current were found to be significant for MRR [119]. Gosai and Joshi, in a study on Si PMEDM of Ti6Al4V, reported that the inclusion of Si powder had a positive effect on both SR and MRR. They reported that high powder concentration, along with a high peak current, could lead to an enhancement in MRR and a reduction in SR [120]. Li et al. employed magnetic stirring in the SiC-mixed EDM of Ti alloy. Their study revealed that the deposited layer of SiC improved surface hardness and the use of high pulse duration could increase the deposited layer strength [121]. By employing the PMEDM process, Prakash et al. were able to generate a nano-porous layer on a β-phase Ti alloy, enhancing its suitability for bone-implant mechanical attachment. This kind of surface property promotes the attachment and proliferation of osteoblastic-like cells [122]. In a 2018 study, Opöz et al. investigated the material deposition phenomena in SiC powder-mixed EDM of Ti-6Al-4V-ELI material. They found that lower currents and higher concentrations improved the deposition, while higher currents reduced it due to fewer secondary discharges [123]. Bui et al. explored the use of PMEDM in antibacterial coating of Ti implants. They incorporated varying concentrations of silver nano-powder into a hydrocarbon-based working medium and machined Ti. The study revealed that the powder concentration significantly influenced the amount of silver deposition [124]. To overcome challenges in regulating silver distribution and spattered layers on surfaces machined and coated with nano-silver PMEDM, Bui et al. introduced ultrasonic vibrations during the machining of Ti-6Al-4V workpiece surfaces. This approach improved silver uniformity and prevented spattering, with the vibration amplitude playing a crucial role in this enhancement [125]. In a comparative study aimed at modifying the surface of medical-grade Ni55.6Ti44.4 alloy, Davis et al. explored the use of copper and brass micro tool and various concentrations (ranging from 0 to 10 g/L) of Zn powder. The findings showed that employing a copper micro tool with 6 g/L powder particle concentrations yielded the most enhanced machinability and surface modification for Ni55.6Ti44.4 alloy. The resultant layer exhibited improved cytocompatibility and increased hydrophobicity on the modified alloy surface [126]. In an experimental study conducted by Shingare and Mujumdar, alumina and SiC powder were introduced to deionized water at three different concentrations. The results highlighted the advantages of PMEDM over conventional EDM, such as improved MRR and surface finish. However, a drawback of the PMEDM process was identified in the form of increased TWR [127]. In an attempt to generate functional surfaces on machined components through the PMEDM process, Venkata Rajesh and Abimannan used MoS_2 micropowder (diameter of 40 μm) and nanopowder (diameter of 90 nm) as powder additives and current as the input. They noted that increasing the current proportionally increased the root mean square roughness for both micro and nano powder-mixed dielectric. At lower current levels, distinct surface characteristics were observed, featuring both negatively skewed and positive kurtosis surfaces in both types of powder-mixed dielectric. The recast layer thickness increased with higher current [128].

Experimental investigations into the various aspects of PMEDM are plenty in the available literature. However, work reported on modeling of the PMEDM process is very limited. Kansal et al. have presented an FEM model for the PMEDM process, but in their work, the total heat distributed to the workpiece was assumed to be a constant value of 9% [129]. Jatti and Bagne presented a thermo-electric model for the PMEDM process, but again the heat distributed to the workpiece was assumed to be a constant [130].

3.6 CHALLENGES AND SCOPE OF FUTURE WORK IN THE FIELD OF PMEDM

Despite the numerous advantages of PMEDM over conventional EDM, PMEDM is still predominantly at a research level. As discussed in Section 3.4.5, there are several challenges associated with implementing PMEDM in industrial applications.

(i) The absence of an efficient filtering method limits the ability to reuse the powder, resulting in increased process costs.
(ii) Sophisticated process control is vital for maintaining a consistent powder suspension, fine-tuning powder concentration, and regulating the influence of powder on spark properties. However, integrating such advanced process control systems for these purposes can pose difficulties and increase expenses.

(iii) In contrast to traditional EDM, the use of powders in the dielectric in PMEDM may lead to higher consumable costs due to the possibility of increased tool wear.
(iv) Effectively managing the disposal or recycling of fine powder particles utilized in PMEDM is imperative, as mishandling can present environmental and health risks. This requires the implementation of costly filtration and extraction systems.
(v) PMEDM might not be suitable for all materials, as some may interact with the powder particles, leading to unwanted deposits during the process. This underscores the importance of carefully selecting the powder, which can be a demanding task.
(vi) Compared to traditional EDM, PMEDM is a relatively new technology. The standardization of process parameters, powder types, and equipment is still evolving, posing challenges for manufacturers in establishing consistent and repeatable production processes.

On account of all these factors, widespread adoption of PMEDM in industries is still challenging. In this context, further research in the following areas can be helpful:

(i) Powder-mixed deionized water is an environmentally friendly dielectric fluid but has limited productivity in terms of MRR. However, hydrocarbon-based dielectric with powder additives can deliver high MRR but is hazardous for the operator and environment. Hence, other alternatives such as rice bran oil and jatropha oil with powder additives should be investigated.
(ii) Dielectric fluids, when used with powder additives, deliver better performance. However, the reusability of the powder is still a concern. Hence ecological and economic feasibility of PMEDM should be investigated with an aim to reuse the powder additives.
(iii) For spark erosion-based machining processes, single-spark simulations are not capable enough to predict the topography of actual machined surfaces. In this regard, realistic multi-spark models would be more useful in simulating the surface topography of actual machined surfaces.
(iv) The carbon adherence phenomenon in the EDM process involving hydrocarbon oil-based dielectrics or graphite PMEDM process adversely affects the predicting capability of models. A detailed study in this direction can improve the accuracy of models.
(v) Additionally, external energy field, viz. magnetic field, can affect the trajectory of motion of debris and in some cases of PMEDM the trajectory of motion of powders (if the powder particles are magnetic). Hence, these factors should be incorporated into the model to make a more accurate prediction of responses.

3.7 CONCLUSION

This chapter successfully presented an overview of EDM and PMEDM processes, covering their applications, advantages, and limitations. In addition to comparing the mechanisms of EDM and PMEDM, this chapter discussed their components and process and performance parameters. It emphasized the importance of modeling of spark erosion-based machining process and highlighted key research findings from previous studies on EDM and PMEDM processes. Furthermore, notable research discoveries related to the PMEDM process have also been presented in this chapter. This chapter has been concluded by outlining the challenges and potential areas for further research in the realm of PMEDM.

REFERENCES

[1] Jeswani ML. Effect of the addition of graphite powder to kerosene used as the dielectric fluid in electrical discharge machining. *Wear* 1981;70:133–139. https://doi.org/10.1016/0043-1648(81)90148-4.

[2] Singh S, Maheshwari S, Pandey PC. Some investigations into the electric discharge machining of hardened tool steel using different electrode materials. *J Mater Process Technol* 2004;149:272–277. https://doi.org/10.1016/j.jmatprotec.2003.11.046.

[3] García Navas V, Ferreres I, Marañón JA, Garcia-Rosales C, Gil Sevillano J. Electro-discharge machining (EDM) versus hard turning and grinding—Comparison of residual stresses and surface integrity generated in AISI O1 tool steel. *J Mater Process* Technol 2008;195:186–194. https://doi.org/10.1016/j.jmatprotec.2007.04.131.

[4] Philip JT, Mathew J, Kuriachen B. Transition from EDM to PMEDM – Impact of suspended particulates in the dielectric on Ti6Al4V and other distinct material surfaces: A review. *J Manuf Process* 2021;64:1105–1142. https://doi.org/10.1016/j.jmapro.2021.01.056.

[5] Volosova MA, Okunkova AA, Povolotskiy DE, Podrabinnik PA. Study of electrical discharge machining for the parts of nuclear industry usage. *Mech Ind* 2015;16:706. https://doi.org/10.1051/meca/2015091.

[6] Ayesta I, Izquierdo B, Sánchez JA, Ramos JM, Plaza S, Pombo I, et al. Influence of EDM parameters on slot machining in C1023 aeronautical alloy. *Procedia CIRP* 2013;6:129–134. https://doi.org/10.1016/j.procir.2013.03.059.

[7] Uhlmann E, Domingos DC. Investigations on vibration-assisted EDM-machining of seal slots in high-temperature resistant materials for turbine components. *Procedia CIRP* 2013;6:71–76. https://doi.org/10.1016/j.procir.2013.03.019.

[8] Ho K., Newman S. State of the art electrical discharge machining (EDM). *Int J Mach Tools Manuf* 2003;43:1287–1300. https://doi.org/10.1016/S0890-6955(03)00162-7.

[9] Kar S, Patowari PK. Electrode wear phenomenon and its compensation in micro electrical discharge milling: A review. *Mater Manuf Process* 2018;33:1491–1517. https://doi.org/10.1080/10426914.2018.1453144.

[10] Dauw DF, Albert L. About the evolution of wire tool performance in wire EDM. *CIRP Ann* 1992;41:221–225. https://doi.org/10.1016/S0007-8506(07)61190-1.

[11] Benedict GF. *Nontraditional manufacturing processes*. Second Ind. New York: Marcel Dekker, Inc.; 2015.

[12] Ganachari VS, Chate UN, Waghmode LY, Mullya SA, Shirguppikar SS, Salgar MM, et al. A comparative performance study of dry and near dry EDM processes in machining of spring steel material. *Mater Today Proc* 2019;18:5247–5257. https://doi.org/10.1016/j.matpr.2019.07.525.

[13] Chakraborty S, Dey V, Ghosh SK. A review on the use of dielectric fluids and their effects in electrical discharge machining characteristics. *Precis Eng* 2015;40:1–6. https://doi.org/10.1016/j.precisioneng.2014.11.003.

[14] Erden A, Bilgin S. Role of impurities in electric discharge machining. *21st Int. Mach. Tool Des. Res. Con*, 1980, pp. 345–350.

[15] Aliyu AA, Abdul-Rani AM, Ginta TL, Prakash C, Axinte E, Razak MA, et al. A review of additive mixed-electric discharge machining: Current status and future perspectives for surface modification of biomedical implants. *Adv Mater Sci Eng* 2017;2017:1–23. https://doi.org/10.1155/2017/8723239.

[16] Talla G, Gangopadhayay S, Biswas C. State of the art in powder-mixed electric discharge machining: A review. *Proc Inst Mech Eng Part B J Eng Manuf* 2017;231:2511–2526. https://doi.org/10.1177/0954405416634265.

[17] Choudhary SK, Jadoun RS. Current advanced research development of electric discharge machining (EDM): A review. *Int J Res Advent Technol* 2014;2:273–297.

[18] Shabgard MR, Badamchizadeh MA, Ranjbary G, Amini K. Fuzzy approach to select machining parameters in electrical discharge machining (EDM) and ultrasonic-assisted EDM processes. *J Manuf Syst* 2013;32:32–39. https://doi.org/10.1016/j.jmsy.2012.09.002.

[19] Chattopadhyay AB, Chattopadhyay AK, Paul S. Manufacturing processes II, IIT Kharagpur_Lec no. 39 n.d.:1–15. https://nptel.ac.in/courses/112105127 (accessed August 31, 2023).

[20] El-Hofy HA-G. *Advanced machining processes*. McGraw-Hill; 2005.

[21] Joshi AY, Joshi AY. A systematic review on powder mixed electrical discharge machining. *Heliyon* 2019;5:e02963. https://doi.org/10.1016/j.heliyon.2019.e02963.

[22] Schulze H-P. *Importance of polarity change in the electrical discharge machining*, 2017, p. 050001. https://doi.org/10.1063/1.5008046.

[23] Wu H, Wang T, Wang J. Research on discharge state detection of finishing in high-speed wire electrical discharge machine. *Int J Adv Manuf Technol* 2019;103:2301–2317. https://doi.org/10.1007/s00170-019-03708-z.

[24] Kunieda M, Lauwers B, Rajurkar KP, Schumacher BM. Advancing EDM through fundamental insight into the process. *CIRP Ann* 2005;54:64–87. https://doi.org/10.1016/S0007-8506(07)60020-1.

[25] Han F, Wachi S, Kunieda M. Improvement of machining characteristics of micro-EDM using transistor type isopulse generator and servo feed control. *Precis Eng* 2004;28:378–385. https://doi.org/10.1016/j.precisioneng.2003.11.005.

[26] Sai Ram J, Jeavudeen S, Mouda PA, Ahamed N. The role of various dielectrics used in EDM process and their environmental, health, and safety issues. *Mater Today Proc* 2023. https://doi.org/10.1016/j.matpr.2023.05.137.

[27] Roy BK, Mandal A. Surface integrity analysis of Nitinol-60 shape memory alloy in WEDM. *Mater Manuf Process* 2019;34:1091–1102. https://doi.org/10.1080/10426914.2019.1628256.

[28] Kumar S, Singh R, Singh TP, Sethi BL. Surface modification by electrical discharge machining: A review. *J Mater Process Technol* 2009;209:3675–3687. https://doi.org/10.1016/j.jmatprotec.2008.09.032.

[29] Jahan MP, Rahman M, Wong YS. Study on the nano-powder-mixed sinking and milling micro-EDM of WC-Co. Int *J Adv Manuf Technol* 2011;53:167–180. https://doi.org/10.1007/s00170-010-2826-9.

[30] Talla G, Gangopadhyay S, Biswas CK. Influence of graphite powder mixed EDM on the surface integrity characteristics of Inconel 625. *Part Sci Technol* 2017;35:219–226. https://doi.org/10.1080/02726351.2016.1150371.

[31] Mohd Abbas N, Solomon DG, Fuad Bahari M. A review on current research trends in electrical discharge machining (EDM). *Int J Mach Tools Manuf* 2007;47:1214–1228. https://doi.org/10.1016/j.ijmachtools.2006.08.026.

[32] Yang X, Guo J, Chen X, Kunieda M. Molecular dynamics simulation of the material removal mechanism in micro-EDM. *Precis Eng* 2011;35:51–57. https://doi.org/10.1016/j.precisioneng.2010.09.005.

[33] Yue X, Yang X. Molecular dynamics simulation of machining properties of polycrystalline copper in electrical discharge machining. *Proc Inst Mech Eng Part B J Eng Manuf* 2019;233:371–380. https://doi.org/10.1177/0954405417748187.

[34] Zhang G, Guo J, Ming W, Huang Y, Shao X, Zhang Z. Study of the machining process of nano-electrical discharge machining based on combined atomistic-continuum modeling method. *Appl Surf Sci* 2014;290:359–367. https://doi.org/10.1016/j.apsusc.2013.11.084.

[35] Wang J, Han F. Simulation model of debris and bubble movement in consecutive-pulse discharge of electrical discharge machining. *Int J Mach Tools Manuf* 2014;77:56–65. https://doi.org/10.1016/j.ijmachtools.2013.10.007.

[36] Huang R, Yi Y, Zhu E, Xiong X. Investigation of a liquid-phase electrode for micro-electro-discharge machining. *Micromachines* 2020;11:935. https://doi.org/10.3390/mi11100935.

[37] Yue X, Yang X, Li Q, Li X. Novel methods for high-speed observation of material removal and molten pool movement in EDM. *Precis Eng* 2020;66:295–305. https://doi.org/10.1016/j.precisioneng.2020.07.009.

[38] Jithin S, Raut A, Bhandarkar U V., Joshi SS. Finite element model for topography prediction of electrical discharge textured surfaces considering multi-discharge phenomenon. *Int J Mech Sci* 2020;177:105604. https://doi.org/10.1016/j.ijmecsci.2020.105604.

[39] Tariq Jilani S, Pandey PC. Analysis and modelling of EDM parameters. *Precis Eng* 1982;4:215–221. https://doi.org/10.1016/0141-6359(82)90011-3.

[40] Pandit SM, Rajurkar KP. A stochastic approach to thermal modeling applied to electro-discharge machining. *J Heat Transfer* 1983;105:555–562. https://doi.org/10.1115/1.3245621.

[41] Ming W, Zhang Z, Wang S, Zhang Y, Shen F, Zhang G. Comparative study of energy efficiency and environmental impact in magnetic field assisted and conventional electrical discharge machining. *J Clean Prod* 2019;214:12–28. https://doi.org/10.1016/j.jclepro.2018.12.231.

[42] Ming W, Zhang G, Li H, Guo J, Zhang Z, Huang Y, et al. A hybrid process model for EDM based on finite-element method and Gaussian process regression. *Int J Adv Manuf Technol* 2014;74:1197–1211. https://doi.org/10.1007/s00170-014-5989-y.

[43] Vignesh M, Ramanujam R. Numerical modelling and experimental validation of crater formation in WEDM hybrid turning of Ti-6Al-4V alloy. P*roc Inst Mech Eng Part E J Process Mech Eng* 2021;235:392–404. https://doi.org/10.1177/0954408920964687.

[44] Ming W, Zhang S, Zhang G, Du J, Ma J, He W, et al. Progress in modeling of electrical discharge machining process. *Int J Heat Mass Transf* 2022;187:122563. https://doi.org/10.1016/j.ijheatmasstransfer.2022.122563.

[45] Dijck FS van, Dutré WL. Heat conduction model for the calculation of the volume of molten metal in electric discharges. *J Phys D Appl Phys* 1974;7:316. https://doi.org/10.1088/0022-3727/7/6/316.

[46] Erden A, Kaftanoglu B. Heat transfer modelling of electric discharge machining. *Proc. Twenty-First Int. Mach. Tool Des. Res. Conf.*, London: Macmillan Education UK; 1981, p. 351–358.

[47] Jithin S, Bhandarkar U V., Joshi SS. Multi-spark model for predicting surface roughness of electrical discharge textured surfaces. *Int J Adv Manuf Technol* 2020;106:3741–3758. https://doi.org/10.1007/s00170-019-04841-5.

[48] Jithin S, Bhandarkar U V., Joshi SS. Analytical simulation of random textures generated in electrical discharge texturing. *J Manuf Sci Eng* 2017;139. https://doi.org/10.1115/1.4037322.

[49] Singh M, Saxena P, Ramkumar J, Rao R V. Multi-spark numerical simulation of the micro-EDM process: an extension of a single-spark numerical study. *Int J Adv Manuf Technol* 2020;108:2701–2715. https://doi.org/10.1007/s00170-020-05566-6.

[50] Nguyen TD, Volgin VM, Lyubimov VV. Simulation of electrical discharge machining of micro-holes, 2020, p. 381–390. https://doi.org/10.1007/978-3-030-22063-1_41.

[51] Kojima A, Natsu W, Kunieda M. Observation of arc plasma expansion and delayed growth of discharge crater in EDM. *Proc. 15th Int. Symp. Electromachining,* ISEM 2007, Pittsburgh: 2007, p. 1–4.

[52] Schumacher BM. About the role of debris in the gap during electrical discharge machining. *CIRP Ann* 1990;39:197–199. https://doi.org/10.1016/S0007-8506(07)61034-8.

[53] Izquierdo B, Sánchez JA, Plaza S, Pombo I, Ortega N. A numerical model of the EDM process considering the effect of multiple discharges. *Int J Mach Tools Manuf* 2009;49:220–229. https://doi.org/10.1016/j.ijmachtools.2008.11.003.

[54] DiBitonto DD, Eubank PT, Patel MR, Barrufet MA. Theoretical models of the electrical discharge machining process. I. A simple cathode erosion model. *J Appl Phys* 1989;66:4095–4103. https://doi.org/10.1063/1.343994.

[55] Snoeys R, Dijck F. Investigation of electro discharge machining operations by means of thermo-mathematical model. *Ann CIRP* 1971;20:35–37.

[56] Beck JV. Transient temperatures in a semi-infinite cylinder heated by a disk heat source. *Int J Heat Mass Transf* 1981;24:1631–1640. https://doi.org/10.1016/0017-9310(81)90071-5.

[57] Beck JV. Large time solutions for temperatures in a semi-infinite body with a disk heat source. *Int J Heat Mass Transf* 1981;24:155–164. https://doi.org/10.1016/0017-9310(81)90104-6.

[58] Tariq Jilani S, Pandey PC. An analysis of surface erosion in electrical discharge machining. *Wear* 1983;84:275–284. https://doi.org/10.1016/0043-1648(83)90269-7.

[59] Pandey PC, Jilani ST. Plasma channel growth and the resolidified layer in edm. *Precis Eng* 1986;8:104–110. https://doi.org/10.1016/0141-6359(86)90093-0.

[60] Patel MR, Barrufet MA, Eubank PT, DiBitonto DD. Theoretical models of the electrical discharge machining process. II. The anode erosion model. *J Appl Phys* 1989;66:4104–4111. https://doi.org/10.1063/1.343995.

[61] Joshi SN, Pande SS. Thermo-physical modeling of die-sinking EDM process. *J Manuf Process* 2010;12:45–56. https://doi.org/10.1016/j.jmapro.2010.02.001.

[62] Bergs T, Schneider S, Harst S, Klink A. Numerical simulation and validation of material loadings during electrical discharge machining. *Procedia CIRP* 2019;82:14–19. https://doi.org/10.1016/j.procir.2019.04.165.

[63] Shabgard M, Ahmadi R, Seyedzavvar M, Oliaei SNB. Mathematical and numerical modeling of the effect of input-parameters on the flushing efficiency of plasma channel in EDM process. *Int J Mach Tools Manuf* 2013;65:79–87. https://doi.org/10.1016/j.ijmachtools.2012.10.004.

[64] Yadav V, Jain VK, Dixit PM. Thermal stresses due to electrical discharge machining. *Int J Mach Tools Manuf* 2002;42:877–888. https://doi.org/10.1016/S0890-6955(02)00029-9.

[65] Singh H. Experimental study of distribution of energy during EDM process for utilization in thermal models. *Int J Heat Mass Transf* 2012;55:5053–5064. https://doi.org/10.1016/j.ijheatmasstransfer.2012.05.004.

[66] Erden A. Effect of materials on the mechanism of electric discharge machining (E.D.M.). *J Eng Mater Technol* 1983;105:132–138. https://doi.org/10.1115/1.3225627.

[67] Zhang Y, Liu Y, Shen Y, Li Z, Ji R, Cai B. A novel method of determining energy distribution and plasma diameter of EDM. *Int J Heat Mass Transf* 2014;75:425–432. https://doi.org/10.1016/j.ijheatmasstransfer.2014.03.082.

[68] Tan PC, Yeo SH. Modelling of overlapping craters in micro-electrical discharge machining. *J Phys D Appl Phys* 2008;41:205302. https://doi.org/10.1088/0022-3727/41/20/205302.

[69] Shabgard M, Ahmadi R, Seyedzavvar M, Oliaei SNB. Mathematical and numerical modeling of the effect of input-parameters on the flushing efficiency of plasma channel in EDM process. *Int J Mach Tools Manuf* 2013;65:79–87. https://doi.org/10.1016/j.ijmachtools.2012.10.004.

[70] Salonitis K, Stournaras A, Stavropoulos P, Chryssolouris G. Thermal modeling of the material removal rate and surface roughness for die-sinking EDM. *Int J Adv Manuf Technol* 2009;40:316–323. https://doi.org/10.1007/s00170-007-1327-y.

[71] Joshi SN, Pande SS. Development of an intelligent process model for EDM. *Int J Adv Manuf Technol* 2009;45:300–317. https://doi.org/10.1007/s00170-009-1972-4.

[72] Philip JT, Mathew J, Kuriachen B. Numerical simulation of the effect of crater morphology for the prediction of surface roughness on electrical discharge textured Ti6Al4V. J *Brazilian Soc Mech Sci Eng* 2020;42:248. https://doi.org/10.1007/s40430-020-02321-6.

[73] Wu H, Ma J, Meng Q, Jahan MP, Alavi F. Numerical modeling of electrical discharge machining of Ti-6Al-4V. *Procedia Manuf* 2018;26:359–371. https://doi.org/10.1016/j.promfg.2018.07.044.

[74] Zahiruddin M, Kunieda M. Comparison of energy and removal efficiencies between micro and macro EDM. *CIRP Ann - Manuf Technol* 2012;61:187–190. https://doi.org/10.1016/j.cirp.2012.03.006.

[75] Zhang R, Zhang Y, Liu Y, Shen Y, Li Z, Wang X. Energy distribution and material removal of electric arc machining (EAM). *J Mater Process Technol* 2017;242:110–116. https://doi.org/10.1016/j.jmatprotec.2016.11.012.

[76] Zhang R, Zhang Y, Liu Y, Shen Y, Li Z, Wang X. Energy distribution and material removal of electric arc machining (EAM). *J Mater Process Technol* 2017;242:110–116. https://doi.org/10.1016/j.jmatprotec.2016.11.012.

[77] Shen Y, Liu Y, Zhang Y, Tan B, Ji R, Cai B, et al. Determining the energy distribution during electric discharge machining of Ti-6Al-4V. *Int J Adv Manuf Technol* 2014;70:11–17. https://doi.org/10.1007/s00170-013-5194-4.

[78] Xia H, Kunieda M, Nishiwaki N. Removal amount difference between anode and cathode in EDM process. *J Japan Soc Electr Mach Eng* 1994;28:31–40. https://doi.org/10.2526/jseme.28.59_31.

[79] Hayakawa S, Yuzawa M, Kunieda M, Nishiwaki N. Time variation and mechanism of determining power distribution in electrodes during EDM process. *Int J Electr Mach* 2001;6:19–25. https://doi.org/10.2526/ijem.6.19.

[80] Singh H. Experimental study of distribution of energy during EDM process for utilization in thermal models. *Int J Heat Mass Transf* 2012;55:5053–5064. https://doi.org/10.1016/j.ijheatmasstransfer.2012.05.004.

[81] DiBitonto DD, Eubank PT, Patel MR, Barrufet MA. Theoretical models of the electrical discharge machining process. I. A simple cathode erosion model. *J Appl Phys* 1989;66:4095–4103. https://doi.org/10.1063/1.343994.

[82] Yeo SH, Kurnia W, Tan PC. Electro-thermal modelling of anode and cathode in micro-EDM. *J Phys D Appl Phys* 2007;40:2513–2521. https://doi.org/10.1088/0022-3727/40/8/015.

[83] Xia H, Hashimoto H, Kunieda M, Nishiwaki N. Measurement of energy distribution in continuous EDM process. *J JSPE* 1996;62:1141–1145.

[84] Zahiruddin M, Kunieda M. *Energy distribution ratio into micro EDM electrodes* * 2014. https://doi.org/10.1299/jamdsm.4.1095.

[85] Xia H, Kunieda M, Nishiwakp N, Lior N. *Measurement of energy distribution into electrodes in EDM processes*. Elsevier Science B.V.; 1994. https://doi.org/10.1016/b978-0-444-81901-7.50114-6.

[86] Okada A, Uno Y, Okajima I. Energy distribution in electrical discharge machining with graphite electrode. *Mem Fac Eng Okayama Univ* 2000;34:19–26.

[87] Jithin S, Raut A, Bhandarkar UV., Joshi SS. FE modeling for single spark in EDM considering plasma flushing efficiency. *Procedia Manuf* 2018;26:617–628. https://doi.org/10.1016/j.promfg.2018.07.072.

[88] Patel MR, Barrufet MA, Eubank PT, DiBitonto DD. Theoretical models of the electrical discharge machining process. II. The anode erosion model. *J Appl Phys* 1989;66:4104–4111. https://doi.org/10.1063/1.343995.

[89] Dijck F Van. *Physico-mathematical analysis of the electro discharge machining process.* Catholic University of Leuven, Belgium, 1973.

[90] Kozak J, Rozenek M, Dabrowski L. Study of electrical discharge machining using powder-suspended working media. *Proc Inst Mech Eng Part B J Eng Manuf* 2003; 217:1597–1602. https://doi.org/10.1243/095440503771909971.

[91] Zhao W, Meng Q, Wang Z. The application of research on powder mixed EDM in rough machining. *J Mater Process Technol* 2002;129:30–33. https://doi.org/10.1016/S0924-0136(02)00570-8.

[92] Tzeng Y-F, Lee C-Y. Effects of powder characteristics on electrodischarge machining efficiency. *Int J Adv Manuf Technol* 2001;17:586–592. https://doi.org/10.1007/s001700170142.

[93] Kansal HK, Singh S, Kumar P. Parametric optimization of powder mixed electrical discharge machining by response surface methodology. *J Mater Process Technol* 2005;169:427–436. https://doi.org/10.1016/j.jmatprotec.2005.03.028.

[94] Kuriachen B, Mathew J. Experimental investigations into the effects of microelectric-discharge milling process parameters on processing Ti–6Al–4V. *Mater Manuf Process* 2015;30:983–990. https://doi.org/10.1080/10426914.2014.984206.

[95] Simao J, Lee H, Aspinwall D, Dewes R, Aspinwall E. Workpiece surface modification using electrical discharge machining. *Int J Mach Tools Manuf* 2003;43:121–128. https://doi.org/10.1016/S0890-6955(02)00187-6.

[96] Kansal HK, Singh S, Kumar P. Technology and research developments in powder mixed electric discharge machining (PMEDM). *J Mater Process Technol* 2007;184:32–41. https://doi.org/10.1016/j.jmatprotec.2006.10.046.

[97] Erden A, Bilgin S. Role of impurities in electric discharge machining. *Proc. Twenty-First Int. Mach. Tool Des. Res. Conf.*, London: Macmillan Education UK; 1981, p. 345–350.

[98] Narumiya H, Mohri N, Saito N, Ootake H, Tsunekawa Y, Takawashi T, et al. EDM by powder suspended working fluid. *Proc 9th Int Symp. Electromachining — ISEM IX,* 1989, p. 5–8.

[99] Kobayashi K, Magara T, Ozaki Y, Yatomi T. The present and future developments of electrical discharge machining. *Proc. 2nd Int. Conf. Die Mould Technol.*, Singapore: 1992, p. 35–47.

[100] Mohri N, Saito N, Narumiya H, Kawatsu Y, Ozaki KT, Kobayashi K. Finish machining by EDM using powder suspended working fluids. *Japan Soc Electr Mach Eng* 1992;49:47–60.

[101] Yu CP, Chen WC, Chang SW, Chang CC. Effects of the concentration of suspended aluminum powder in dielectric fluid on EDM of carbide of tungsten. *Proc. 13th Conf. Chinese Soc. Mech. Eng.*, Taiwan: 1996, p. 445–450.

[102] Ming QY, He LY. Powder-suspension dielectric fluid for EDM. *J Mater Process Technol* 1995;52:44–54. https://doi.org/10.1016/0924-0136(94)01442-4.

[103] Wong Y., Lim L., Rahuman I, Tee W. Near-mirror-finish phenomenon in EDM using powder-mixed dielectric. *J Mater Process Technol* 1998;79:30–40. https://doi.org/10.1016/S0924-0136(97)00450-0.

[104] Uno Y, Okada A, Yamada T, Hayashi Y, Tabuchi Y. Surface integrity in EDM of aluminum bronze with nickel powder mixed fluid. *J Japan Soc Electr Mach Eng* 1998;32:24–31. https://doi.org/10.2526/jseme.32.70_24.

[105] Chow H-M, Yan B-H, Huang F-Y, Hung J-C. Study of added powder in kerosene for the micro-slit machining of titanium alloy using electro-discharge machining. *J Mater Process Technol* 2000;101:95–103. https://doi.org/10.1016/S0924-0136(99)00458-6.

[106] Furutania K, Saneto A, Takezawa H, Mohri N, Miyake H. Accretion of titanium carbide by electrical discharge machining with powder suspended in working fluid. *Precis Eng* 2001;25:138–144. https://doi.org/10.1016/S0141-6359(00)00068-4.

[107] Peças P, Henriques E. Influence of silicon powder-mixed dielectric on conventional electrical discharge machining. *Int J Mach Tools Manuf* 2003;43:1465–1471. https://doi.org/10.1016/S0890-6955(03)00169-X.
[108] Klocke F, Lung D, Antonoglou G, Thomaidis D. The effects of powder suspended dielectrics on the thermal influenced zone by electrodischarge machining with small discharge energies. *J Mater Process Technol* 2004;149:191–197. https://doi.org/10.1016/j.jmatprotec.2003.10.036.
[109] Wu KL, Yan BH, Huang FY, Chen SC. Improvement of surface finish on SKD steel using electro-discharge machining with aluminum and surfactant added dielectric. *Int J Mach Tools Manuf* 2005;45:1195–1201. https://doi.org/10.1016/j.ijmachtools.2004.12.005.
[110] Kansal HK, Singh S, Kumar P. An experimental study of the machining parameters in powder mixed electric discharge machining of Al 10%SiCP metal matrix composites. *Int J Mach Mater* 2006;1:396. https://doi.org/10.1504/IJMMM.2006.012349.
[111] Yeo SH, Tan PC, Kurnia W. Effects of powder additives suspended in dielectric on crater characteristics for micro electrical discharge machining. *J Micromechanics Microengineering* 2007;17:N91–N98. https://doi.org/10.1088/0960-1317/17/11/N01.
[112] Chow H-M, Yang L-D, Lin C-T, Chen Y-F. The use of SiC powder in water as dielectric for micro-slit EDM machining. *J Mater Process Technol* 2008;195:160–170. https://doi.org/10.1016/j.jmatprotec.2007.04.130.
[113] Wu KL, Yan BH, Lee J-W, Ding CG. Study on the characteristics of electrical discharge machining using dielectric with surfactant. *J Mater Process Technol* 2009;209:3783–3789. https://doi.org/10.1016/j.jmatprotec.2008.09.005.
[114] Kumar S, Singh R. Investigating surface properties of OHNS die steel after electrical discharge machining with manganese powder mixed in the dielectric. *Int J Adv Manuf Technol* 2010;50:625–633. https://doi.org/10.1007/s00170-010-2536-3.
[115] Ojha K, Garg RK, Singh K. Experimental investigation and modeling of PMEDM process with chromium powder suspended dielectric. *Int J Appl Sci Eng* 2011;9:65–81.
[116] Syed KH, Palaniyandi K. Performance of electrical discharge machining using aluminium powder suspended distilled water. *Turkish J Eng Env Sci* 2012;36:195–207. https://doi.org/10.3906/muh-1202-2.
[117] Bhattacharya A, Batish A, Kumar N. Surface characterization and material migration during surface modification of die steels with silicon, graphite and tungsten powder in EDM process. *J Mech Sci Technol* 2013;27:133–140. https://doi.org/10.1007/s12206-012-0883-8.
[118] Goyal S, Singh RK. Parametric study of powder mixed EDM and optimization of MRR & surface roughness. *Int J Sci Eng Technol* 2014;3:56–62.
[119] Kumar H. Development of mirror like surface characteristics using nano powder mixed electric discharge machining (NPMEDM). *Int J Adv Manuf Technol* 2015;76:105–113. https://doi.org/10.1007/s00170-014-5965-6.
[120] Gosai ND, Joshi AY. Experimental investigation and optimization of process parameters used in the silicon powder mixed electro discharge machining of Ti-6Al-4V alloy using response surface methodology. *J Manuf Sci Prod* 2016;16:21–32. https://doi.org/10.1515/jmsp-2015-0013.
[121] Li L, Zhao L, Li ZY, Feng L, Bai X. Surface characteristics of Ti-6Al-4V by SiC abrasive-mixed EDM with magnetic stirring. *Mater Manuf Process* 2017;32:83–86. https://doi.org/10.1080/10426914.2016.1151043.
[122] Prakash C, Kansal HK, Pabla BS, Puri S. On the influence of nanoporous layer fabricated by PMEDM on β -Ti implant: Biological and computational evaluation of bone-implant interface. *Mater Today Proc* 2017;4:2298–2307. https://doi.org/10.1016/j.matpr.2017.02.078.

[123] Öpöz TT, Yaşar H, Ekmekci N, Ekmekci B. Particle migration and surface modification on Ti6Al4V in SiC powder mixed electrical discharge machining. *J Manuf Process* 2018;31:744–758. https://doi.org/10.1016/j.jmapro.2018.01.002.
[124] Bui VD, Mwangi JW, Schubert A. Powder mixed electrical discharge machining for antibacterial coating on titanium implant surfaces. *J Manuf Process* 2019;44:261–270. https://doi.org/10.1016/j.jmapro.2019.05.032.
[125] Bui VD, Boehme V, Mwangi JW, Berger T, Schubert A. Investigation of ultrasonic vibration assisted powder mixed electrical discharge machining for antibacterial coating on implant surfaces. *Procedia CIRP* 2020;95:431–436. https://doi.org/10.1016/j.procir.2020.01.159.
[126] Davis R, Singh A, Debnath K, Sabino RM, Popat K, da Silva LRR, et al. Surface modification of medical-grade Ni55.6Ti44.4 alloy via enhanced machining characteristics of Zn Powder Mixed-µ-EDM. *Surf Coatings Technol* 2021;425:127725. https://doi.org/10.1016/j.surfcoat.2021.127725.
[127] Shingare PP, Mujumdar S. Experimental investigation of powder-mixed dielectrics for electrical discharge machining. *Manuf Lett* 2022;33:428–436. https://doi.org/10.1016/j.mfglet.2022.07.056.
[128] Venkata Rajesh J, Abimannan G. Powder mixed electrical discharge machining: One step method to create functional surfaces. *Mater Lett* 2023;351:135103. https://doi.org/10.1016/j.matlet.2023.135103.
[129] Kansal HK, Singh S, Kumar P. Numerical simulation of powder mixed electric discharge machining (PMEDM) using finite element method. *Math Comput Model* 2008;47:1217–1237. https://doi.org/10.1016/j.mcm.2007.05.016.
[130] Jatti VS, Bagane S. Thermo-electric modelling, simulation and experimental validation of powder mixed electric discharge machining (PMEDM) of BeCu alloys. *Alexandria Eng* J 2018;57:643–653. https://doi.org/10.1016/j.aej.2017.02.023.

4 Machine Learning-Based Predictive Modeling of the EDM Process

Shahid Ali and Asma Perveen

ABBREVIATIONS

CO	Cutting speed override
CR	cutting rate
DD	Dimensional deviation
DT	Dwell time
El	Electrode material
EWR	Electrode wear rate
GP	Gap voltage
I_p	Peak Current
LTMS	Layer thickness of mild steel
LTSS	Later thickness of stainless steel
MRR	Material removal rate
OC	Overcut
O_r	Orientation
PO	Profile offset
P_r	Pressure ratio
SR	Surface roughness
t	machining time
T_{off}	Pulse off time
T_{on}	Pulse on time
TWR	Tool wear rate
V	Spark Voltage
WD	Wire distance
WF	Wire feed
WP	workpiece material
WEDM	Wire electrodischarge machining

4.1 INTRODUCTION

Machine learning (ML) is the study that empowers computers to learn and refine from experience without being categorically programmed (Mackenzie 2015). ML

DOI: 10.1201/9781003769675-4

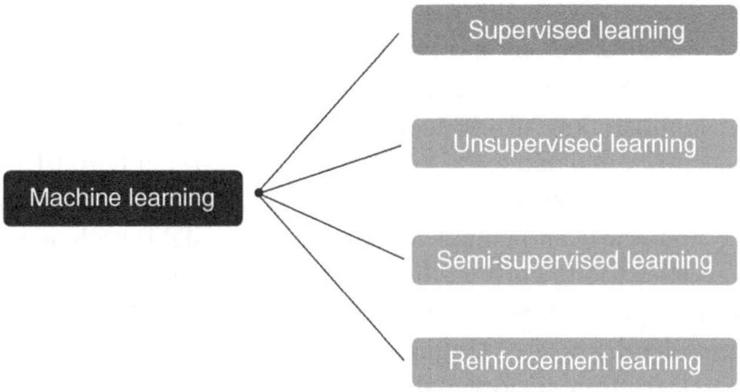

FIGURE 4.1 Types of machine learning.

focuses on teaching machines to handle data more efficiently. Often, interpreting and extracting valuable information from data can be challenging. In such cases, ML algorithms are employed to facilitate this process. ML employs various algorithms to address diverse data problems. Data scientists insist that there is no universal algorithm appropriate for all problems. The preference of the algorithm pivots on the nature of the problem, the number of variables associated, and the type of model (Richards and Zill 2019; Lee et al. 2021; Lee 2023). Figure 4.1 depicts the types of learning.

4.1.1 Importance of Non-Conventional Machining

The requirement of modern industries led to the widespread adoption of hard materials and geometrically intricate components with superior mechanical, thermal, and tribological properties. Machining techniques for such materials are mainly divided into traditional and nontraditional machining methods (Zadafiya et al. 2021). Although traditional machining can achieve position accuracies and tolerances, traditional mechanical machining encounters specific challenges with these emerging materials and intricate geometries, for instance very low material removal rate (MRR) and poor surface integrity (Zhao et al. 2024).

In contrast, nontraditional machining methods, particularly electro-discharge machining (EDM), provide a novel solution for adequately and precisely machining hard materials and parts, especially in the aerospace and biomedical industries (Korgal et al. 2024). This technique lowers cutting forces and temperatures, prolongs tool life, and permits the use of larger machining parameters. As a result, EDM significantly enhances overall machining performance, making it a crucial technique for advanced manufacturing applications.

4.1.2 Why Do We Need ML for EDM?

Researchers ponder in the realm of EDM to predict surface topography and machining parameters through numerical modeling and simulation. Numerous analytical and numerical models, such as two-dimensional (2D) and three-dimensional (3D) models, have been suggested in the literature to simulate surface profiles (Izquierdo et al. 2011; Feng et al. 2016; Jithin and Joshi 2021). Traditional numerical models, while valuable, face limitations in capturing the full complexity of EDM. These models often require simplifying assumptions (homogeneity of the workpiece, incompressible and laminar flow of the melt, negligible radiation effects, etc.) that can hinder their accuracy and applicability. The inherent variability in EDM stems from factors like tool wear, debris removal, and fluid dynamics. In contrast, ML excels in managing the non-linear and complex interactions of EDM, such as those between machining variables (pulse current, discharge voltage, pulse on time, pulse off time, etc.), performance parameters, and material properties (MRR, crater size, surface roughness, microhardness, etc.) (Jithin and Joshi 2021). The data-driven approach of ML allows it to learn from historical data and adapt to new conditions without extensive re-calibration, resulting in more precise predictions of performance parameters based on machining conditions (Oyebode et al. 2023). ML models can continuously improve as they are exposed to more data, making them particularly well-suited for processes with high variability. Additionally, ML models can provide real-time predictions with minimal computational overhead, unlike numerical models that are often computationally intensive, expensive, and time-consuming. The following sections include the integration of ML in the domain of EDM modeling and optimization. The primary objective of this chapter is to provide a comprehensive overview of ML optimization algorithms. Each algorithm is examined regarding its functionality, operational principles, and EDM applications. Practical applications of these algorithms are showcased to demonstrate how ML can solve complex, multi-objective optimization problems and improve system efficiencies, particularly in handling non-linear interactions and high-dimensional input spaces common in EDM processes.

4.2 MACHINE LEARNING MODELS AND EDM

4.2.1 Artificial Neural Network (ANN)

Artificial neural networks (ANNs) are highly parallel computational models that mimic the function of the human brain. An ANN consists of numerous simple processors or "neurons," linked by weighted connections as depicted in Figure 4.2. The output of each neuron depends solely on the locally available information, either stored internally or received via these weighted connections. Each neuron collects inputs from multiple other neurons and conveys its output to additional neurons.

In ANNs, the input data to a processing element, denoted as i_i , are increased by connection weights (w_i), simulating the reinforcement of neural pathways in the brain. Learning in ANNs occurs through the adjustment of these weights. The weight-adjusted input values are aggregated using a vector-to-scalar function, such as summation $\left(y = \sum w_i x_i \right)$, averaging, maximum input, or mode value, to generate a single input value for the neuron. The processing element uses a transfer function to generate its output after calculating the input value. This output is then used as the input for the neu7rons in the following layer. The transfer function uses a sigmoid, hyperbolic tangent or other non-linear functions to change the input value of

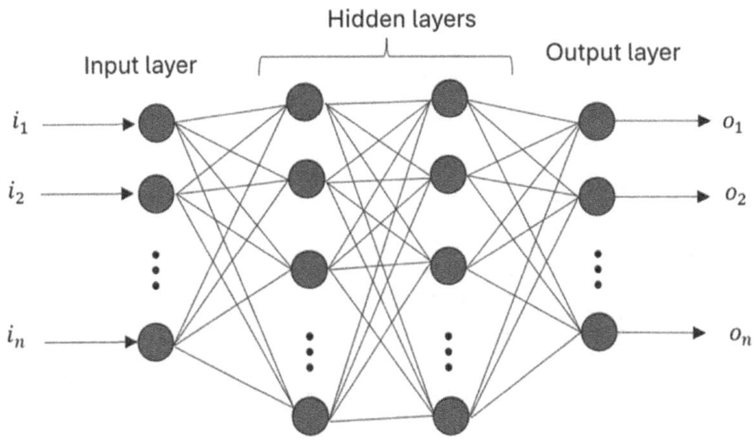

FIGURE 4.2 Schematic of ANN model.

a neuron. The neural network generates a final output value, o_i, or a vector of values by repeating this process over numerous layers of processing units (Schmidgall et al. 2024).

Electro-discharge machining (EDM) is a sophisticated machining technique used to machine hard and electrically conductive materials by employing electrical discharges. Despite its advantages in precision machining, EDM's complexity poses challenges in predicting process outcomes due to the non-linear interactions among various parameters such as discharge current, pulse duration, electrode material, and dielectric fluid properties (Jithin and Joshi 2021). To address these challenges, many researchers have turned to ANN to predict the responses associated with EDM. Table 4.1 outlines the responses attained by researchers using ANN.

4.2.1.1 Multi-Objective Genetic Algorithm (MOGA)

The Multi-Objective Genetic Algorithm (MOGA) is an advanced evolutionary technique used to optimize problems involving multiple conflicting objectives. MOGA works by searching for various solutions using genetic operations like mutation and crossover. The evaluated solution for each trial, according to its performance on the given objectives, and the best solutions are chosen for subsequent generations. These selected solutions are then recombined to produce new, potentially superior solutions, with the process iterating until an optimal or satisfactory solution is found. The schematic of the process flow is depicted in Figure 4.3. For instance, Sana et al. (2024a) employed an ANN with three to ten variable neurons in the hidden layer and observed the better-fit data with R^2 greater than 0.98 for all response measures. Based on the objectives to minimize the electrode wear rate (EWR) and overcut (OC), the MOGA predicts the optimized response better than the experiment design (DOE). The confirmatory results show that, with non-treated electrodes, EWR and OC are 67.02% and 59.73% better than the DOE results, respectively. In contrast, the corresponding improvements obtained using cryogenically treated electrodes are higher, reaching 78.41% for EWR and 67.79% for OC.

TABLE 4.1
Application of ANN in the domain of EDM

Author	Type of EDM	Material	Process variables	Performance variables	Optimization technique	R-square	Findings
(Sana et al. 2024a)	PMEDM	Aluminum Al6061	I_p, V, T_{on}, C_p	EWR, OC	ANN with Multi-Objective Genetic Algorithm (MOGA)	>0.98	Employing MOGA results in 65.02% and 59.73% in EWR and OC by using non-treated brass electrodes. Similarly, by using cryogenically treated electrodes, the improvement in EWR and OC is 78.41% and 67.79%.
(Sana et al. 2024b)	PMEDM	Aluminum Al6061	I_p, V, T_{on}, C_p	MRR, SR, SEC	ANN with NSGA-II	>0.98	Employing NSGA-II results in an improvement in MRR, SR, and SEC by 87.42%, 3.4%, and 0.7%, respectively.
(Jafarian 2020)	EDM	Inconel 718	I_p, T_{on}, V, T_{off}	MRR, SR	ANN with NSGA-II		The average error for the MRR and SR is 6.9% and 7.8%, respectively
(Hurairah et al. 2024)	PMEDM	Stainless steel SS316	I_p, V, T_{on}, T_{off}	Radial Overcut (Rd_OC)	ANN with Mult objective genetic algorithm (MOGA)	>0.99	The author reported 40.21% and 34.37% improvement in radial OC for brass and aluminum electrodes, respectively, by employing MOGA optimization.
(Quarto et al. 2021)	u-EDM	Stainless steel AISI304	I_p, V, F, D, WP, El	MRR, t, DD	ANN-PSO		ANN-PSO predicts MRR and DD 20% better than the FEM simulation
(Ishfaq et al. 2023b)	EDM	Inconel 617 (IN617)	dielectrics, surfactants, type of electrode, treatment type	OC	Levenberg–Marquart algorithm	>0.96	They observed that the hidden layer with four neurons shows comparatively better results than with five neurons; also, it can predict the OC better within the designed space.

(Continued)

TABLE 4.1
(Continued)

Author	Type of EDM	Material	Process variables	Performance variables	Optimization technique	*R*-square	Findings
(Ishfaq et al. 2023)	WEDM	stainless clad steel	LTSS, LTMS, V, D, P_r, f, T_{on}, O_r	Cutting speed	ANN, SVM, ELM		The ANN ($R^2 = 0.9$) predicts the cutting speed response much better than the SVM ($R^2 = 0.36$), and ELM ($R^2 = 0.88$) models
(Manoj et al. 2023)	WEDM	Pyromet-680	WD, DT, PO, CO	CR, SR	ANN-BP, ANFIS	>0.991	The average percent error is much less by using ANFIS (0%–5%) than ANN (0%–10%)
(Pourasl et al. 2022)	EDM	AISI-D6 steel	I_p, V, T_{on}	MRR, TWR, SR	ANN, ANFIS		The root mean square error for the MRR, TWR, and SR (0.81, 0.28, and 0.17) predicted by the ANFIS design is better than the ANN

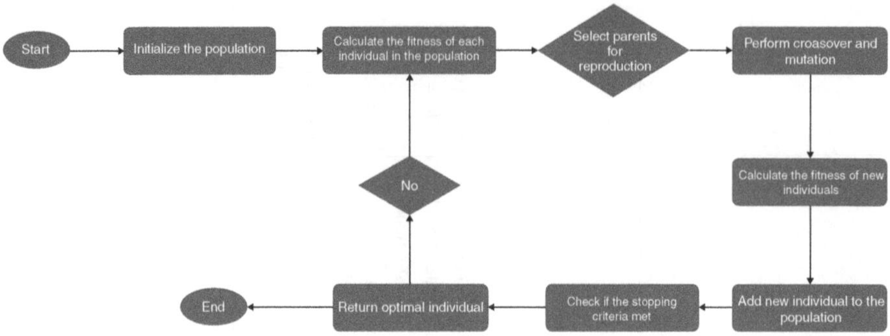

FIGURE 4.3 The schematic diagram for the MOGA process flow (Sana et al. 2024a).

4.2.1.2 Non-Dominated Sorting Genetic Algorithm (NSGA)-II

The Non-Dominated Sorting Genetic Algorithm (NSGA)-II is an optimization method designed to address multi-objective optimization problems. It operates by iteratively generating and evolving a population of potential solutions. Through the process of non-dominated sorting and crowding distance assignment, NSGA-II effectively identifies a diverse set of trade-off solutions along the Pareto front. This method guides the search toward superior Pareto front solutions over successive iterations. The key steps for NSGA-II are as follows:

- First, the initial population of the NSGA-II algorithm is selected as per the dimensionality of the variables.
- After evaluating the function of individuals at each iteration, the individuals are categorized based on their fitness and crowding distances.
- Then the top-ranked individuals were chosen as parents, and genetic operators generated offspring individuals.
- Lastly, the parents and offspring were integrated, and the non-dominated sorting process was performed to evaluate their fitness.

The schematic of the process flow is illustrated in Figure 4.4.

For instance, Jafarian (2020) employed the NSGA-II algorithm to determine the optimized input parameters for better MRR and SR. He reported the minimum mean squared error with one hidden layer and 15 and 12 neurons in the ANN function for MRR and SR, respectively. The study reported that with the Logsig-Transig function, the mean testing and mean training error were less than 6%. By applying the optimization algorithm, the errors with the experimental results were less than 8% for both MRR and SR.

4.2.1.3 Particle Swarm Optimization (PSO)

Another evolutionary technique is Particle Swarm Optimization (PSO), which uses a group of particles in the pursuit space to optimize a fitness function. Particles are first arranged at random in the quest for space, and they assess their grade or fitness at those locations. Every particle relocates to an alternate

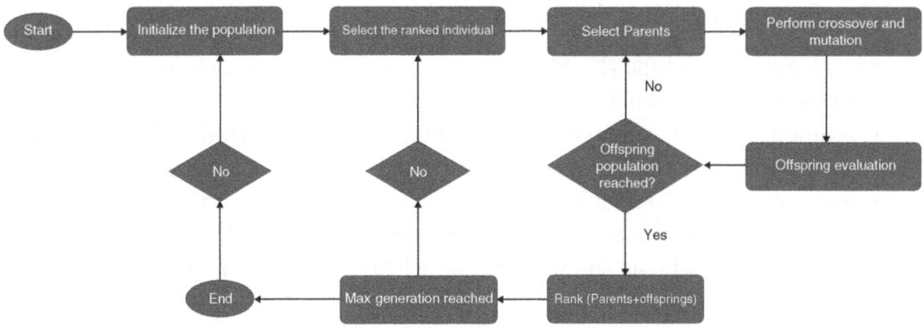

FIGURE 4.4 The schematic diagram of the process flow for NSGA-II (Chang et al. 2015).

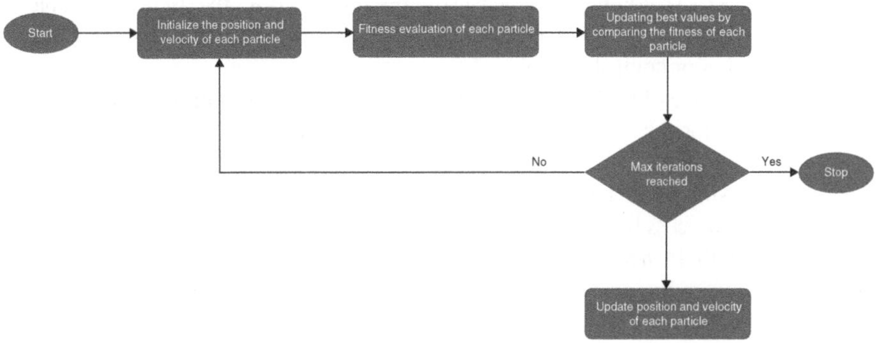

FIGURE 4.5 Schematic of the PSO algorithm (van Zyl and Engelbrecht 2023).

spot that provides greater fitness than its prior position for a predefined number of iterations. Together with random disturbances, this motion is dictated by the particle's optimal position history and the most favorable locations that other particles in the swarm have reached. The swarm as a whole converges toward the best solution that achieves the fitness function in the problem space by means of these iterative updates. Since PSO lacks a selection operation, all particles in the swarm are kept throughout the search, in contrast to other evolutionary methods. Every iteration updates each particle's location and velocity according to its best position as well as the best positions the group has yet to achieve. This updated mechanism ensures that each particle explores the search space by adjusting its trajectory toward better solutions. The schematic of the algorithm is depicted in Figure 4.5.

For instance, Quarto et al. (2021) employed the ANN integrated with PSO to optimize the MRR, dimensional deviation, and t with six input parameters (I_p, V, F, D, WP (workpiece material), El (electrode material)). They concluded that the implementation of the ANN-PSO algorithm can more accurate predictions than the finite element method (FEM). The error between the ANN-PSO forecasted and experimental results was less than 30%; whereas the FEM-based predictions exhibited an average error of approximately 50% when estimating MRR and DD.

4.2.2 Extreme Learning Machine

Extreme learning machine (ELM) is a popular learning approach for single hidden layer feedforward neural networks used in classification and regression applications. ELM is unique in that it can randomly assign input weights and hidden layer biases, as well as adaptively alter the number of hidden layer nodes. The least squares approach is used to compute the output layer weights, completing the learning process with a single, iteration-free mathematical transformation (Ding et al. 2015). However, along with the advantage of being simple and easy, ELM has significant drawbacks (Rajput et al. 2024).

- The foundation of ELM is empirical risk minimization, which ignores structural risk and can result in overfitting.
- Because ELM computes the least squares solution directly, users are unable to adjust the model to account for the unique features of the datasets, which leads to poor controllability.
- The presence of outliers in the datasets can have a substantial impact on the ELM performance, indicating a lack of robustness.

For instance, Ishfaq et al. (2023a) employed the ELM model to predict the cutting speed of the WEDM process based on eight input variables. The authors reported that an ELM architecture with 15 neurons in the hidden layer achieved the best predictive performance, yielding an R^2 value of approximately 0.95 along with RSME = 0.02 mm/s.

4.2.3 Support Vector Machine (SVM)

An artificial intelligence (AI) modeling approach called Support Vector Machine (SVM) is used to create models that correlate input and output variables. It is frequently used to classify and approximate complex and non-linear data. SVM uses hyperplanes to divide input data into two different classes. As depicted in Figure 4.6, these hyperplanes are composed of a positive and a negative hyperplane as well as an extra line known as the maximum margin hyperplane. There is a specific distance between each data point in both the negative or positive hyperplane and the margin. SVM models typically perform better on datasets that are farther away from the maximum margin hyperplane in terms of convergence and generalization. By creating an effective SVM model, the structural risk minimization (SRM) approach seeks to lower the response variable's absolute limit of prediction error.

For instance, Ishfaq et al. (2023a) integrated the SVM model to predict the cutting speed of the WEDM process based on eight input response parameters. By employing the grid search algorithm, they have observed the optimum combination of hyperplane values where the model has an $R^2 = 0.87$ with RSME = 0.04 mm/s.

4.2.4 Convolutional Neural Network (CNN)

Conventional feature extraction and deep learning-based methods are the two main approaches for image-based regression and classification. Convolutional neural

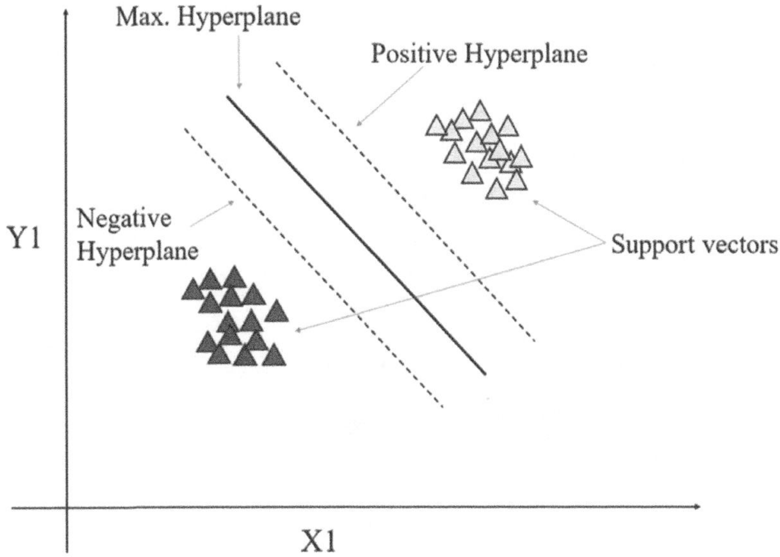

FIGURE 4.6 Schematic of SVM model (Ishfaq et al. 2023a).

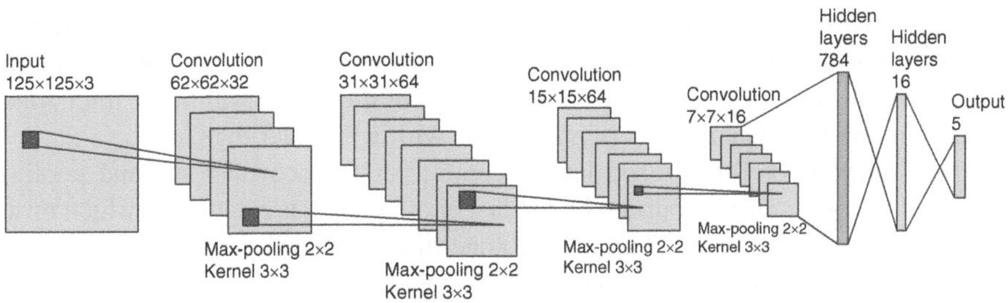

FIGURE 4.7 Schematic structure of CNN (Phung and Rhee 2019).

networks (CNNs) are a well-known deep learning model that is intended for processing data with a grid pattern, like photographs. A typical CNN has three different types of layers: convolutional, pooling, and fully connected layers, as shown in Figure 4.7. The convolutional and pooling layers are in charge of feature extraction, and the fully connected layer uses these features to produce the final output, like classification.

An essential part of CNNs is the convolutional layer, which performs a number of mathematical operations including convolution, a particular type of linear operation. By applying several filters to the input data, this layer aids in the detection of different features in the input image. In contrast, the pooling layer lowers the dimensionality of the feature maps, which simplifies the network and lessens its computing burden. Lastly, the input image is classified into numerous groups by the fully connected layer using the high-level characteristics that the convolutional and pooling layers have extracted (Dey et al. 2023).

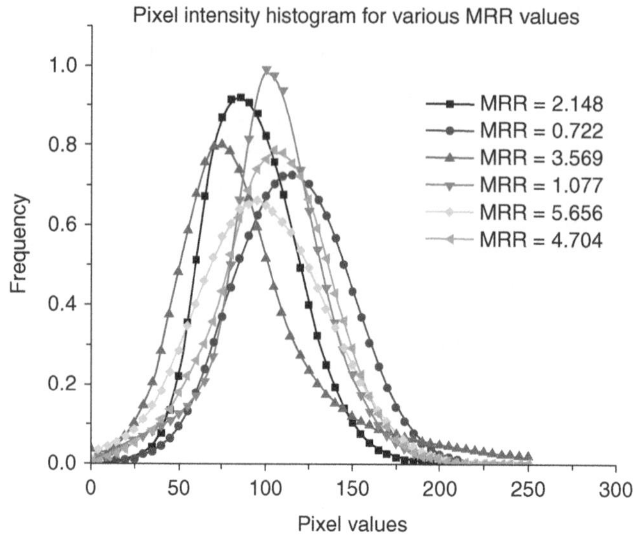

FIGURE 4.8 Pixel intensity of numerous MRR (Balaji and Narendranath 2023).

Filters slide across picture maps in the CNN's pooling layer to accomplish subsampling, which shrinks picture map sizes while maintaining important characteristics. Highest pooling and average pooling are the two popular varieties of pooling layers. While average pooling yields the average of the values within the filter zone, maximum pooling yields the maximum value within the filter region.

Dense layers, or fully linked layers, come after the convolutional and pooling layers. The image maps produced by the last pooling layer are flattened, which turns them from multi-dimensional shapes into one-dimensional shapes, and then concatenated into multi-tuple vectors in order to connect to the dense layer. In these deep layers, the dropout mechanism is frequently used to prevent over-fitting. Prior to getting the output layer, the multi-tuple vector must first transit across zero, one, or more dense layers. The softmax function is commonly employed in the output layer of CNNs to provide a probability distribution across the potential classes when the CNN's goal is to categorize the input picture (Li et al. 2022).

For instance, Balaji and Narendranath (2023) conducted the CNN-based scanning electron microscopic SEM image classification after the wire-EDM process. They divided the data sample into six classes based on the input parameters (pulse on time (T_{on}), pulse off time (T_{off}), servo voltage (SV), wire feed (WF)). The data (pixel statistics) was extracted from the pre-processed SEM images and rounded to the multiple of 5 in the range of 0–255, as shown in Figure 4.8. The CNN architecture includes the convolutional 1-D layer, max pooling layer, and two dense layers. The prediction accuracy for the MRR was above 81% for all classes.

4.3 SUMMARY

EDM is a stochastic process, so the involvement of many variables (such as T_{on}, T_{off}, flushing, variation in temperature, non-homogeneity of tool and workpiece) makes it challenging to model the performance parameters numerically. So, the previous researchers implemented ML algorithms to forecast the performance parameters. In this chapter, we looked at several ML approaches and how they were used in predicting various performance parameters, focusing on EDM modeling and optimization. The goal of ML algorithms like NSGA-II and MOGA is to find balanced trade-off solutions in order to address multi-objective optimization problems. To improve fitness function optimization, PSO models the cooperative movement of particles. ANNs learn by adjusting weights and simulating brain functions. In contrast, ELMs fast adapt single hidden layer feedforward neural networks without repeated training. With an emphasis on reducing structural risk, SVMs divide data into classes using hyperplanes in order to perform classification and regression tasks. CNNs classify grid-pattern data, such as photographs, by applying the softmax function through a series of convolutional, pooling, and fully connected layers. These algorithms are highly effective at handling intricate, non-linear interactions, which makes them especially well-suited for highly variable processes. This increases their robustness and efficiency across a wide range of applications.

REFERENCES

Balaji V, Narendranath S. 2023. Optimization of wire-EDM process parameters for Ni–Ti-Hf shape memory alloy through particle swarm optimization and CNN-based SEM-image classification. *Results Eng.* 18. https://doi.org/10.1016/j.rineng.2023.101141

Chang Y, Bouzarkouna Z, Devegowda D. 2015. Multi-objective optimization for rapid and robust optimal oilfield development under geological uncertainty. *Comput Geosci.* 19(4):933–950. https://doi.org/10.1007/s10596-015-9507-6

Dey K, Kalita K, Chakraborty S. 2023. Prediction performance analysis of neural network models for an electrical discharge turning process. *Int J Interact Des Manuf.* 17(2):827–845. https://doi.org/10.1007/s12008-022-01003-y

Ding S, Zhao H, Zhang Y, Xu X, Nie R. 2015. Extreme learning machine: Algorithm, theory and applications. *Artif Intell Rev.* 44(1):103–115. https://doi.org/10.1007/s10462-013-9405-z

Feng X, Wong YS, Hong GS. 2016. Characterization and geometric modeling of single and overlapping craters in micro-EDM. *Mach Sci Technol.* 20(1):79–98. https://doi.org/10.1080/10910344.2015.1085317

Hurairah MA, Sana M, Farooq MU, Anwar S. 2024. Genetic algorithm-based optimization of artificial neural network of process parameters and characterization of machining errors in graphene mixed dielectric. *Arab J Sci Eng.* 49(11):15649–15666. https://doi.org/10.1007/s13369-024-09029-y

Ishfaq K, Sana M, Ashraf WM. 2023a. Artificial intelligence–built analysis framework for the manufacturing sector: Performance optimization of wire electric discharge machining system. *Int J Adv Manuf Technol.* 128(11–12):5025–5039. https://doi.org/10.1007/s00170-023-12191-6

Ishfaq K, Sana M, Waseem MU, Ashraf WM, Anwar S, Krzywanski J. 2023b. Enhancing EDM machining precision through deep cryogenically treated electrodes and ANN modelling approach. *Micromachines (Basel).* 14(8). https://doi.org/10.3390/mi14081536

Izquierdo B, Sánchez JA, Ortega N, Plaza S, Pombo I. 2011. Insight into fundamental aspects of the EDM process using multidischarge numerical simulation. *Int J Adv Manuf Technol.* 52(1–4):195–206. https://doi.org/10.1007/s00170-010-2709-0

Jafarian F. 2020. Electro discharge machining of Inconel 718 alloy and process optimization. *Mater Manuf Process.* 35(1):95–103. https://doi.org/10.1080/10426914.2020.1711919

Jithin S, Joshi SS. 2021. Surface topography generation and simulation in electrical discharge texturing: A review. *J Mater Process Technol.* 298. https://doi.org/10.1016/j.jmatprotec.2021.117297

Korgal A, Shettigar AK, Karanth PN, Prabhakar DAP. 2024. Advances in micro electro discharge machining of biomaterials: A review on processes, industrial applications, and current challenges. *Mach Sci Technol.* 28(2):215–265. https://doi.org/10.1080/10910344.2024.2311376

Lee WH. 2023. The choice of machine learning algorithms impacts the association between brain-predicted age difference and cognitive function. *Mathematics.* 11(5). https://doi.org/10.3390/math11051229

Lee WH, Antoniades M, Schnack HG, Kahn RS, Frangou S. 2021. Brain age prediction in schizophrenia: Does the choice of machine learning algorithm matter? *Psychiatry Res Neuroimaging*. 310. https://doi.org/10.1016/j.pscychresns.2021.111270

Li Z, Liu F, Yang W, Peng S, Zhou J. 2022. A survey of convolutional neural networks: Analysis, applications, and prospects. *IEEE Trans Neural Netw Learn Syst.* 33(12):6999–7019. https://doi.org/10.1109/TNNLS.2021.3084827

Mackenzie A. 2015. The production of prediction: What does machine learning want? *Eur J Cult Stud.* 18(4–5):429–445. https://doi.org/10.1177/1367549415577384

Manoj I V., Manjaiah M, Narendranath S. 2023. Optimization and prediction of responses using artificial neural network and adaptive neuro-fuzzy interference system during taper profiling on pyromet-680 using wire electric discharge machining. *J Mater Eng Perform.* 32(3):993–1005. https://doi.org/10.1007/s11665-022-07165-w

Oyebode O, Fowles J, Steeves D, Orji R. 2023. Machine learning techniques in adaptive and personalized systems for health and wellness. *Int J Hum Comput Interact.* 39(9):1938–1962. https://doi.org/10.1080/10447318.2022.2089085

Phung VH, Rhee EJ. 2019. A High-accuracy model average ensemble of convolutional neural networks for classification of cloud image patches on small datasets. *Appl. Sci. (Switz.).* 9(21). https://doi.org/10.3390/app9214500

Pourasl HH, Javidani M, Khojastehnezhad VM, Vatankhah Barenji R. 2022. The performance prediction of electrical discharge machining of AISI D6 tool steel using ANN and ANFIS techniques: A comparative study. *Crystals (Basel).* 12(3). https://doi.org/10.3390/cryst12030343

Quarto M, D'urso G, Giardini C, Maccarini G, Carminati M. 2021. A comparison between finite element model (FEM) simulation and an integrated artificial neural network (ANN)-particle swarm optimization (PSO) approach to forecast performances of micro electro discharge machining (micro-EDM) drilling. *Micromachines (Basel).* 12(6). https://doi.org/10.3390/mi12060667

Rajput BS, Roy P, Soni S, Raghuwanshi BS. 2024. ELM-based imbalanced data classification-A review. *Informatica (Slovenia).* 48(2):185–204. https://doi.org/10.31449/inf.v48i2.5082

Richards MJ, Zill JC. 2019. *Australasian Transport Research Forum.* [place unknown].

Sana M, Asad M, Farooq MU, Anwar S, Talha M. 2024a. Machine learning for multi-dimensional performance optimization and predictive modelling of nanopowder-mixed electric discharge machining (EDM). *Int J Adv Manuf Technol.* 130(11–12):5641–5664. https://doi.org/10.1007/s00170-024-13023-x

Sana M, Asad M, Farooq MU, Anwar S, Talha M. 2024b. Sustainable electric discharge machining using alumina-mixed deionized water as dielectric: Process modelling by artificial neural networks underpinning net-zero from industry. *J Clean Prod.* 441. https://doi.org/10.1016/j.jclepro.2024.140926

Schmidgall S, Ziaei R, Achterberg J, Kirsch L, Hajiseyedrazi SP, Eshraghian J. 2024. Brain-inspired learning in artificial neural networks: A review. *APL Machine Learning.* 2(2). https://doi.org/10.1063/5.0186054

Zadafiya K, Dinbandhu, Kumari S, Chattarjee S, Abhishek K. 2021. Recent trends in non-traditional machining of shape memory alloys (SMAs): A review. *CIRP J Manuf Sci Technol*. 32:217–227. https://doi.org/10.1016/j.cirpj.2021.01.003

Zhao G, Zhao B, Ding W, Xin L, Nian Z, Peng J, He N, Xu J. 2024. Nontraditional energy-assisted mechanical machining of difficult-to-cut materials and components in aerospace community: A comparative analysis. *Int J Extreme Manuf*. 6(2). https://doi.org/10.1088/2631-7990/ad16d6

van Zyl JP, Engelbrecht AP. 2023. Set-based particle swarm optimisation: A review. *Mathematics*. 11(13). https://doi.org/10.3390/math11132980

5 Electric Discharge Machining of Ceramics and Composites

Shanmuga Priyan V.G. and Kanmani Subbu S.

5.1 INTRODUCTION

Ceramics are a class of engineering materials that have been used by human beings for millennia. Some of the earliest man-made objects were ceramic-based like earthen vessels or pots. The word Ceramics is derived from the Greek word 'keramikos' meaning 'burnt stuff', referring to the process in which the ceramic mixture is heated to a very high temperature to harden or set it. Earthenware, stoneware, porcelain, bricks, and tiles are the common applications of ceramics. These ceramics were known as the traditional ceramics and are mostly made by firing clay, sometimes mixed with Silica. The bonding in ceramics can range from purely ionic to totally covalent or a combination of both. It is usually a compound of metals and non-metals. This type of bonding is the reason for the characteristic properties of ceramics. They generally have a very high melting temperature, high creep resistance, high thermal resistance, high chemical resistance, high Young's modulus, strength, and hardness even at elevated temperatures. Though they exhibit the above-said desirable properties, they also show poor ductility and fracture toughness. Due to ionic bonding, the availability of free electrons to conduct electricity is very low. So traditional ceramics have very less electrical conductivity. The high melting temperatures and thermal resistance of ceramics make them suitable candidates for refractory bricks and linings. Due to all the above-said properties, traditional ceramics are used in applications where metals or other engineering materials cannot be employed. Ceramics are broadly classified into oxide (e.g. zirconia, alumina) and non-oxide ceramics (e.g. silicon carbide, silicon nitride). Most of the ceramics are abrasive in nature and are used in many unconventional machining processes as cutting tool material. Advanced ceramics are basically modified ceramics and ceramic composites that have enhanced properties in comparison to the traditional ceramics. Electroceramics are a class of advanced ceramics known as semi-metallic ceramics, which have modified electrical conductivity. They are extensively used in modern capacitors, resistors, modulators, substrates for integrated circuits, etc.[1]. Advanced ceramics also exhibit magnetic and optical properties. Ceramic fibers, whiskers, and particles are commonly used reinforcements in composites.

DOI: 10.1201/9781003769675-5

Composites are engineered materials of very high demand that are used in almost all fields of life. Composites are fastly replacing conventional materials owing to their desirable properties like high strength-to-weight ratio and stiffness-to-weight ratio. A composite consists of two or more constituents that retain their distinct physical and chemical identities during fabrication and service. The important characteristic of composites is their tailorable properties. Composites have properties that are unique from the individual constituents. Composites generally consist of a continuous phase – matrix and a discontinuous phase – reinforcement. The primary function of the reinforcement is load bearing, and it gives the desired physical and mechanical properties to the composite. Matrix protects the reinforcement from the environment, helps in stress distribution, and holds the reinforcement in position. These materials started to be used in aerospace and automobile applications where the above-said properties were key requirements and slowly found their way in other applications such as sporting goods, casings, marine structures, ballistic armors, and brake pads/drums. Composites are classified based on the matrix material into polymer matrix composites (PMCs), metal matrix composites (MMCs), and ceramic matrix composites (CMCs). Each class of composites has its own set of advantages and disadvantages. MMCs have better fire resistance, higher temperature operation range, resistance to chemical attack and moisture than PMCs, and better ductility over CMCs. CMCs have better impact toughness than both PMCs and MMCs and can also be operated at very high temperatures and corrosive environments. Based on the type of reinforcement used, composites are classified into fiber-reinforced composites and particulate-reinforced composites. Fiber-reinforced composites are further classified into short fiber- and long fiber-reinforced composites. Fiber-reinforced composites are anisotropic, whereas whisker-reinforced and particle-reinforced composites are isotropic. Structural composites are those that have high bending stiffness and low weight. Honeycomb structures and laminate composites are examples of structural composites[2]. Despite being materials of importance, the widespread use of ceramics and composites is limited by the difficulties in fabrication and processing.

5.2 LIMITATIONS IN EXISTING MACHINING METHODS

5.2.1 Machining of Ceramics

Due to the very high brittleness, low impact toughness, and poor thermal conductivity of ceramics, they are difficult to be machined by conventional machining processes. The very low thermal conductivity of these materials also causes thermal shocks with the usage of conventional coolants. Sintering is the commonly used fabrication process for ceramics, and sintered ceramics have very high hardness. So, machining is done to some extent in the pre-sintered state or green condition, and this is known as Green Machining. Polycrystalline diamond (PCD) tools, tungsten carbide, and high-speed steel tools coated with titanium nitride are used as tool materials in green machining[3].

Grinding is the major machining process applied to sintered ceramics. A diamond of various grit sizes bonded with resin is used as the grinding wheel. The feed during grinding is kept very low to avoid the formation of microcracks, which

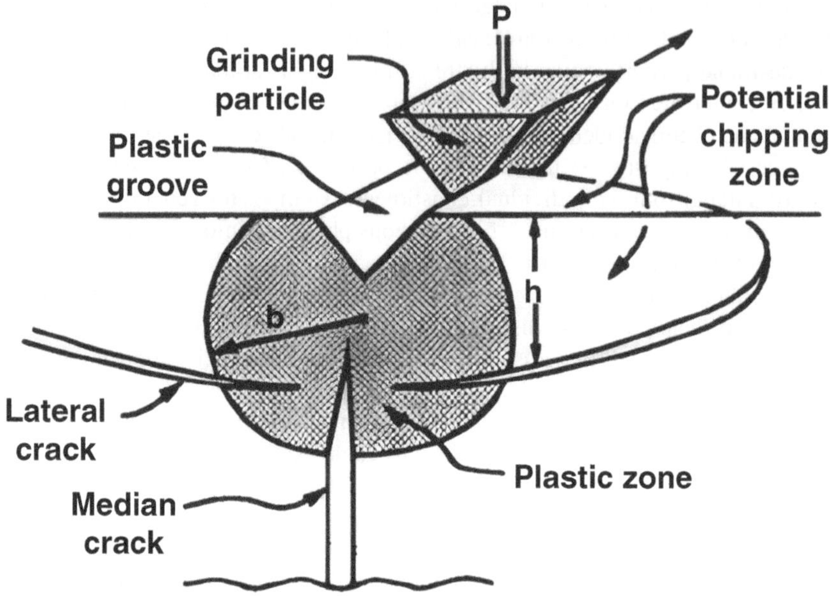

FIGURE 5.1 Indentation fracture mechanism of grinding of ceramics[4].

will eventually lead to failure of the material. The material removal in grinding is hypothesized to follow the two mechanisms: (i) indentation fracture mechanism and (ii) machining approach. Figure 5.1 shows the lateral and median cracks that are generally observed in the grinding of ceramics by the indentation fracture mechanism. Extensive studies done on the indentation fracture mode showed that grain dislodgement and intergranular micro-fracture are the dominant modes of failure, especially at elevated temperatures[3–5].

Machining mode of material removal happens in grinding when the depth of cut is minimum or lesser than a critical depth. For any brittle material, a ductile regime of material removal happens within the critical depth. This is usually in the range of 1–3.5 μm for all ceramic materials. Figure 5.2 shows the scratching in Alumina with a diamond tool, which resembles the machining mode of material removal for various depths of cut. For a depth of cut less than 1 μm, ductile flow with plowed ridges was observed, which implies that the load applied is lesser than the nominal load required to develop lateral cracks (Figure 5.2a). For 1–3.5 μm depth of cut, scale-like cracks and plastic flow were observed, and the material was removed in the form of fine particles, which still comes in the ductile regime (Figure 5.2b). For a depth of cut greater than 10 μm, the material failed by chipping, which is characteristic of brittle failure (Figure 5.2c)[4].

The material removal rate (MRR) in grinding is very low, and the cost involved in removing stock material is high compared to other machining processes. To overcome these issues, advanced machining processes were developed. Laser beam machining (LBM), ultrasonic machining (USM), abrasive water jet machining (AWJM), and laser-assisted machining are some of the successfully used advanced machining processes for ceramic processing.

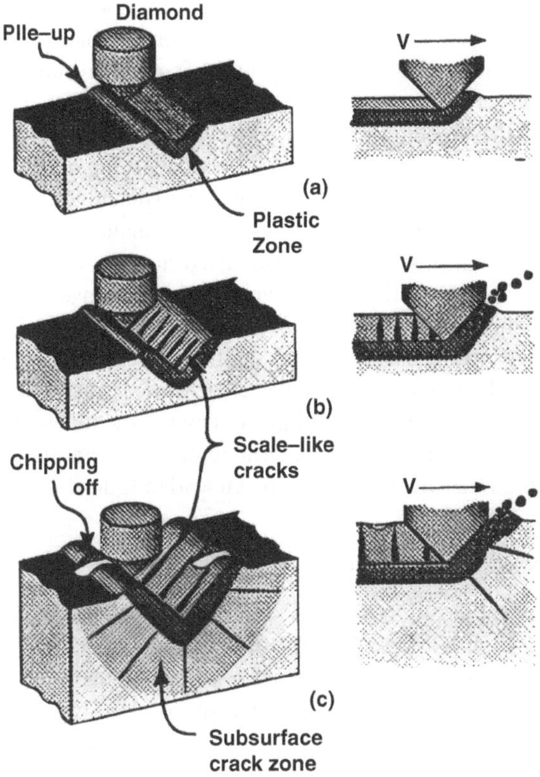

FIGURE 5.2 Scratching of alumina showing machining mode of material removal in ceramics for (a) depth of cut < 1 μm, (b) depth of cut 1–3.5 μm, (c) depth of cut > 10 μm[4].

In LBM, a highly focused laser beam with very high energy density is used to melt and vaporize the material in the workpiece. LBM shows a high MRR but decreases the strength of the material in many cases due to thermal damage and crack formation[6]. Tsai et al. came up with a method to machine ceramics by laser milling, grooves were made around the machining zone using a laser beam, and a defocused laser was used to irradiate and heat the workpiece. Due to the very low thermal conductivity of the ceramics, it experiences thermal shocks, and the material is removed by fracture. The energy efficiency of this process is lower, and the proper focusing and control of the beam and other parameters are very difficult[7].

Ultrasonic machining (USM) uses an abrasive slurry between an ultrasonically vibrating tool and the stationary workpiece for material removal. The material is removed by the chipping action of the abrasives in the slurry. In addition to the vibrating tool USM, rotary ultrasonic machining (RUM) is also commonly used. Though these processes are widely used, MRR and the aspect ratio that can be achieved are less and the tool wear rate (TWR) is very high due to the inherent hardness of ceramics[8].

Abrasive water jet machining (AWJM) uses a very high velocity water jet containing fine abrasive particles to machine the workpiece. The material is removed by erosion action caused by the kinetic energy of the abrasive particles along with the water jet[9]. Jiyue Zeng et al. studied the erosion mechanism of polycrystalline ceramic materials. Intergranular cracking due to fractures resulting from the impact stress waves of the abrasive particle jet caused material removal. The machining rate is very low in AWJM, so ceramics cannot be economically machined. Stray cutting and kerf width irregularities are observed in the machining of thicker sections[10].

Hybrid machining processes are also being used to machine ceramics. They utilize the combined advantages of both processes and eliminate the disadvantages associated with the individual processes. Laser-assisted machining is a hybrid process in which a laser is used along with other conventional or unconventional processes. It is a common term which may involve a wide range of processes like laser-assisted turning, laser-assisted milling, laser-assisted drilling, laser-assisted electric discharge machining (LAEDM), and laser-assisted electrochemical machining (LAECM). The laser beam employed heats the material just in front of the tool and softens the material. This helps in the ductile mode of material removal in ceramics. But laser-assisted processes cause high thermal loads, which can lead to failure of ceramic materials.

5.2.2 Machining of Composites

Composites have enhanced properties compared to their parent materials, and they usually have high hardness, which limits their machinability by conventional machining processes. PMCs generally have a lower melting point, and it is difficult to machine by conventional machining processes as effectively as metals and their

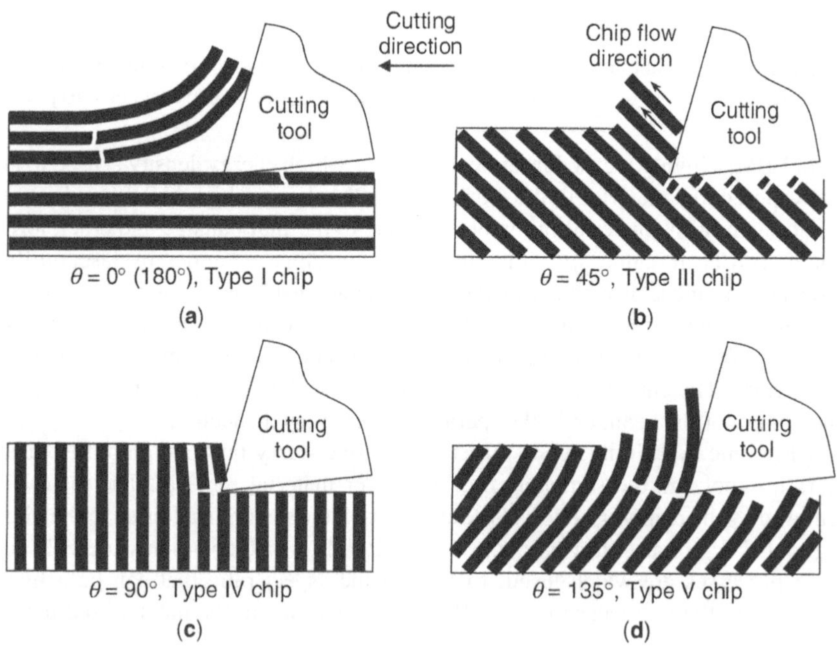

FIGURE 5.3 Chip formation modes in machining of unidirectional composites[11].

alloys. Conventional machining processes such as turning, end-milling, and drilling can all be performed on PMCs, but each has its own limitations. The temperature produced during machining is kept to a minimum as possible by using a sharp tool with good thermal conductivity, high speeds, low feed rates, and low depths of cut. Figure 5.3 shows the types of chips formed for different fiber orientations for a unidirectional PMC. The orientation of the fibers in the composite is the major factor deciding the efficiency of the machining process. If the fibers are oriented perpendicular to the cutting direction, the tool wear will be less, and for all other angles it is high. Fiber pull-out is common in unidirectional composites, where due to the cutting forces the fibers fracture along their length and get removed as semi-continuous chips[11].

Delamination is the most common problem faced in the drilling of multilayered PMCs. It is the separation of layers in the composite layup. Unconventional machining processes most widely used for PMCs are AWJM and LBM. AWJM is beneficial for cutting or machining thin sections, but in machining thicker sections, the deviations in the jet can impregnate the plies or the layers and cause delamination[12]. Laser beam cutting increases the temperature of the material at localized spots, and this affects the properties of the material, causing heat-affected zones (HAZs), and chars the material.

The commonly used reinforcements in MMCs are ceramic particles and whiskers, which have very high strength and hardness, even higher than the tool material. This causes rapid tool wear and abrasion of the cutting edge. Due to the inhomogeneous nature of composites, the tool would come in contact with the softer matrix and harder reinforcement frequently, which causes failure of the tool. During conventional machining of MMCs, the reinforcement particles or fibers used either get cut through or get pulled out, which affects the surface quality and integrity of the MMC. Figure 5.4 shows the reinforcement particles being cut through or pulled out during the machining of particle-reinforced MMCs[13].

Built-up edge formation is a predominant problem in the machining of MMCs, especially at low cutting speeds. These not only reduce the cutting forces at lower

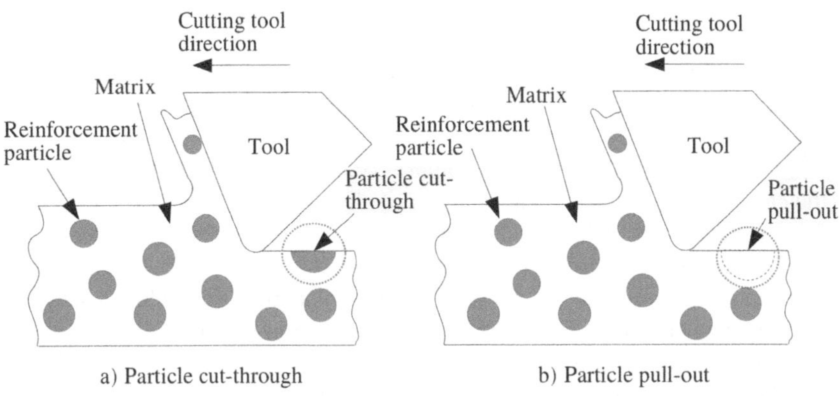

FIGURE 5.4 Schematic showing the machining of particle-reinforced MMCs. (a) Particle cut-through and (b) particle pull-out[13].

cutting speeds but also affect the surface finish, geometric accuracy, and the rake angle of the tool.

Usually, during the machining of metals and alloys, increasing the cutting speed gives a better surface finish, but that is not the case with MMC machining. The higher cutting speed increases the temperature at the chip-tool interface, and the temperature of the tool increases, which softens the tool and increases the tool wear. PCD tools show better performance than other tools. But PCD tools are expensive compared to other tools, which further adds to the cost involved in the machining of MMCs[14].

CMCs show better fracture toughness than their ceramic parent materials. The reinforcements are usually in the form of whiskers or fine powders. In some cases, in addition to increasing the fracture toughness, reinforcements also increase the electrical and thermal conductivity of the ceramic composite. Such composites can be machined in the presence of a coolant without the thermal shock problems as observed in ceramics. All the machining processes discussed for ceramics are suitable for CMCs, and similar limitations are faced.

The conventional machining of ceramics and composites is uneconomical and fails to meet the tight tolerances and geometric accuracies due to the issues already discussed. Advanced machining processes are being extensively used for the machining of such materials.

5.3 ELECTRIC DISCHARGE MACHINING AS A SOLUTION

Electric discharge machining (EDM) is a process that is gaining potential and widespread usage in the machining of advanced engineering materials. It can be effectively used to machine any electrically conductive material, irrespective of its hardness. The capabilities and applications of the spark erosion process are multifold. EDM is used to machine complex and intricate 3D features, holes with aspect ratio higher than 100, make an array of minute holes and even cutting with wire-cut EDM (WEDM).

5.3.1 Principle

The material removal in EDM is due to the electrothermal energy of the controlled sparks produced between the workpiece and the tool electrodes in the presence of a dielectric. Figure 5.5 shows a schematic of a typical die sinking-type electro discharge machine with reverse polarity configuration. The tool is a replica of the cavity to be produced. The spark is produced in the narrowest gap between the two electrodes, which is decided by the surface roughness of the electrodes. The inter-electrode gap should be very small, and it is usually in the range of a few micrometers and is maintained by servo control. The dielectric helps in setting up a conductive channel and sustains the spark. The commonly used dielectric media are hydrocarbon oils like kerosene, transformer oil, EDM oil, or deionized water. When the potential difference applied between the two electrodes is higher than the breakdown voltage of the dielectric, it gets ionized and sets up a conductive path, which produces a spark[15,16].

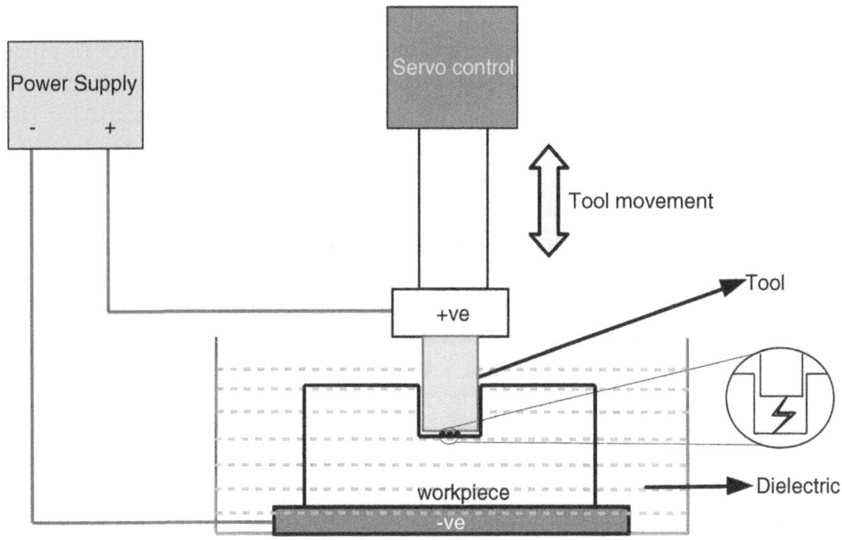

FIGURE 5.5 Schematic representation of a typical EDM system.

5.3.2 Material Removal Mechanism

The spark produced soon develops into a plasma channel with very high temperatures which melts and vaporizes the material. The material is removed from the surface of the electrodes by the formation of craters. The molten metal and the vapors get condensed by the dielectric and form debris. This debris is removed by the circulation of the dielectric. So, the dielectric helps in producing the spark, removing the debris, and also cooling the electrodes.

The molten metal also solidifies over the surface of the workpiece and is known as a recast layer. This layer is undesirable and affects the surface finish and properties of the material. There are many variants in the EDM process, and the commonly used ones are die sinking-type EDM and WEDM. In WEDM, a thin wire of electrically conductive material is used as the tool. It is widely used in cutting and shaping a wide range of materials. Complex and intricate shapes can be machined effectively with EDM and its variants[17].

5.3.3 Process Parameters

Material removal is the ultimate objective in any machining process, so MRR becomes an important response or output. Certain applications require the product material to have a certain surface finish for efficient functioning and surface integrity. Hence, surface finish is the second most important output. In EDM, not only does the workpiece lose material due to sparks, but the tool does too. So, parameters like tool wear ratio, relative wear ratio, and electrode wear rate are also important. Geometric accuracies, form accuracies, tolerances, etc., decide the efficacy and essentially the suitability of a process for machining. These outputs or responses are affected by machining conditions, workpiece and tool properties, and other factors which are commonly referred to as process parameters. Understanding the effect of process parameters on the responses becomes necessary to fully understand a process.

Some of the important process parameters in EDM are as follows:

a) Current
The current is applied between the electrodes by the power generator. Peak current refers to the maximum current that is achieved in each pulse. Average current is the current achieved over a cycle. It is the product of duty cycle and peak current. Current starts flowing between the electrodes after the open gap voltage has reached a maximum value necessary to form an ionization path in the dielectric. It then reaches the maximum preset value or the peak current.
b) Voltage
It is the potential difference applied between the electrodes, i.e. the tool and workpiece. It includes open gap voltage and working gap voltage. Breakdown voltage is the voltage necessary to start ionization in the dielectric. Before the current starts to flow, the open gap voltage increases to a maximum and reduces to the working gap voltage once the current starts flowing.
c) Pulse on-time and pulse off-time
Pulse on-time is the duration during which active sparking and material removal happen. The MRR depends on the energy available during the pulse on-time. Pulse off-time is the pause in the operation of EDM; no material removal takes place. With the start of pulse off-time, the spark disassociates, and the deionization of the dielectric and flushing of the used dielectric happen. A sufficient pulse off-time is necessary for the stability of machining. Pulse on-time and off-time together constitute the cycle time.
d) Frequency
It is the number of cycles that occur in the inter-electrode gap for one second. The higher the frequency, the lesser will be the on-time and subsequent material removal, so the surface finish increases. Low frequency results in higher material removal and is suitable for roughing operations. The duration of the sparks produced is very short, in the range of a few microseconds, and the frequency of the sparks is very high, up to 10^6 sparks per second.
e) Duty cycle
The duty cycle is the ratio of pulse on-time to the total cycle time. It gives a measure of the efficiency of the process and is expressed in percentage. The higher the duty cycle, the higher will be the pulse on-time and hence material removal.
f) Polarity
There are two polarities possible in EDM, depending on the nature of the power terminal to which the electrodes are connected. Straight polarity is the one in which the tool is connected to the negative terminal or becomes the cathode, and the workpiece is the anode. In this case, the material removal from the workpiece will be more and is suitable for a roughing

operation. In reverse polarity, the tool is connected to the positive terminal and becomes the anode, and the workpiece becomes the cathode. In this case, the tool wear will be more and material removal from the workpiece will be less, hence suitable for finishing operations and micro-EDM.

g) Spark gap
It is the minimum gap between the electrodes for which the spark occurs. If the spark gap is more, the flushing will be more, but the ionization path and spark will not be formed. This parameter is not usually considered because it is servo-controlled. However, to find out the minimum spark gap for a given energy setting, the spark gap can be set manually and studied.

5.4 ELECTRIC DISCHARGE MACHINING OF CERAMICS

Electric discharge machining can be used to effectively machine ceramics with higher MRR and accuracy than that in grinding[18]. The tools used in EDM are ductile materials like copper or brass, so the high tool material cost involved in green machining is eliminated.

5.4.1 Conducting and Insulating Ceramics

One major limitation of EDM is that it is generally suitable only for electrically conductive materials. Materials with electrical resistivity less than 100 Ω cm can be readily machined with EDM. Most ceramic materials are inherently insulative, i.e. have very high electrical resistivity. Very few ceramics like titanium diboride (TiB_2) and boron carbide (B_4C) and semi-metallic ceramics like lead oxide (PbO), ruthenium dioxide (RuO_2), bismuth ruthenate ($Bi_2Ru_2O_7$), and bismuth iridate ($Bi_2Ir_2O_7$) have appreciable electrical conductivity and can be machined as they are by the EDM process. Semi-metallic ceramics are commonly called advanced electroceramics[19].

5.4.2 Assisting Electrode Method

A method of EDM suitable for insulating ceramics known as the assisting electrode method (AEM) came into existence a few decades ago. In AEM, a conductive material is coated over the insulating ceramic, and initially, the discharge happens between the tool and this assisting electrode in the dielectric medium. Figure 5.6 shows a schematic of a typical AEM system with the conductive layer and machining action explained. As the assisting material erodes away, the carbon particles in the hydrocarbon dielectric used form a conductive layer on the workpiece, and subsequent discharges take place[20].

A researcher attempted EDM of zirconia (ZrO_2) by using a conductive lacquer-based material as the assisting electrode. The conductive lacquer consists of carbon black suspended in polymethyl methacrylate (PMMA). The coating was done by two methods, such as the doctor knife method and screen printing. The thickness and electrical conductivity of the coatings were characterized before conducting the structuring, and cylindrical copper tungsten (WCu) and copper electrodes of 1 mm

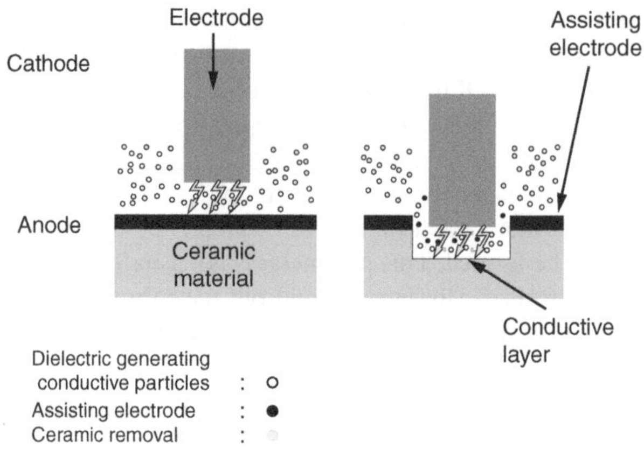

FIGURE 5.6 Schematic representation of a typical AEM system[20].

and 0.5 mm diameter, respectively, were used as the tool materials[20]. The presence of conductive debris and conductive particles from the cracked dielectric oil sustains the process. Copper mesh plates were used over Sialon ceramics to machine them with copper as the tool material. Si_3N_4, SiC, ZrO_2, Al_2O_3, and glass were also successfully machined by AEM[21]. The addition of graphite powders to the dielectric along with the assisting electrode also proved to be an effective method. EDM of glass, ZrO_2, and Al_2O_3 was successfully carried out by this method, and satisfactory results were obtained[22].

5.4.3 Material Removal Mechanism

Melting and evaporation are the commonly observed material removal mechanisms in electrically conductive materials in EDM. In some studies, it is noted that ceramic materials show mechanical erosion caused by the sudden contraction and expansion during EDM. This is considered to be due to thermal spalling since the thermal conductivity of ceramic materials is very low.

Figure 5.7 shows the surface of ceramics machined using the assisting electrode method of EDM. In glass, the material removal is primarily by mechanical erosion (Figure 5.7a). The ZrO_2 surface showed smaller and equally distributed craters (Figure 5.7b). SiC showed a better surface finish compared to all other ceramic materials (Figure 5.7e). Since B_4C has sufficient electrical conductivity, it can readily be machined by EDM. The machined surfaces of B_4C and TiO_2 show material removal mostly by thermal modes (Figures 5.7c and 5.7d). ZrO_2 with Y_2O_3 addition also showed material removal similar to that of ZrO_2 (Figure 5.7f). Ceramic materials with very less electrical and thermal conductivity show higher tendencies for material removal by spalling or mechanical erosion[22].

Pure polycrystalline TiB_2 showed material removal in the form of flakes. Once a spark disintegrates, the very high temperature at the spot is suddenly reduced by the dielectric, which is similar to quenching. The diffusion of heat into the material produces conflicting compressive and tensile stresses parallel and perpendicular to

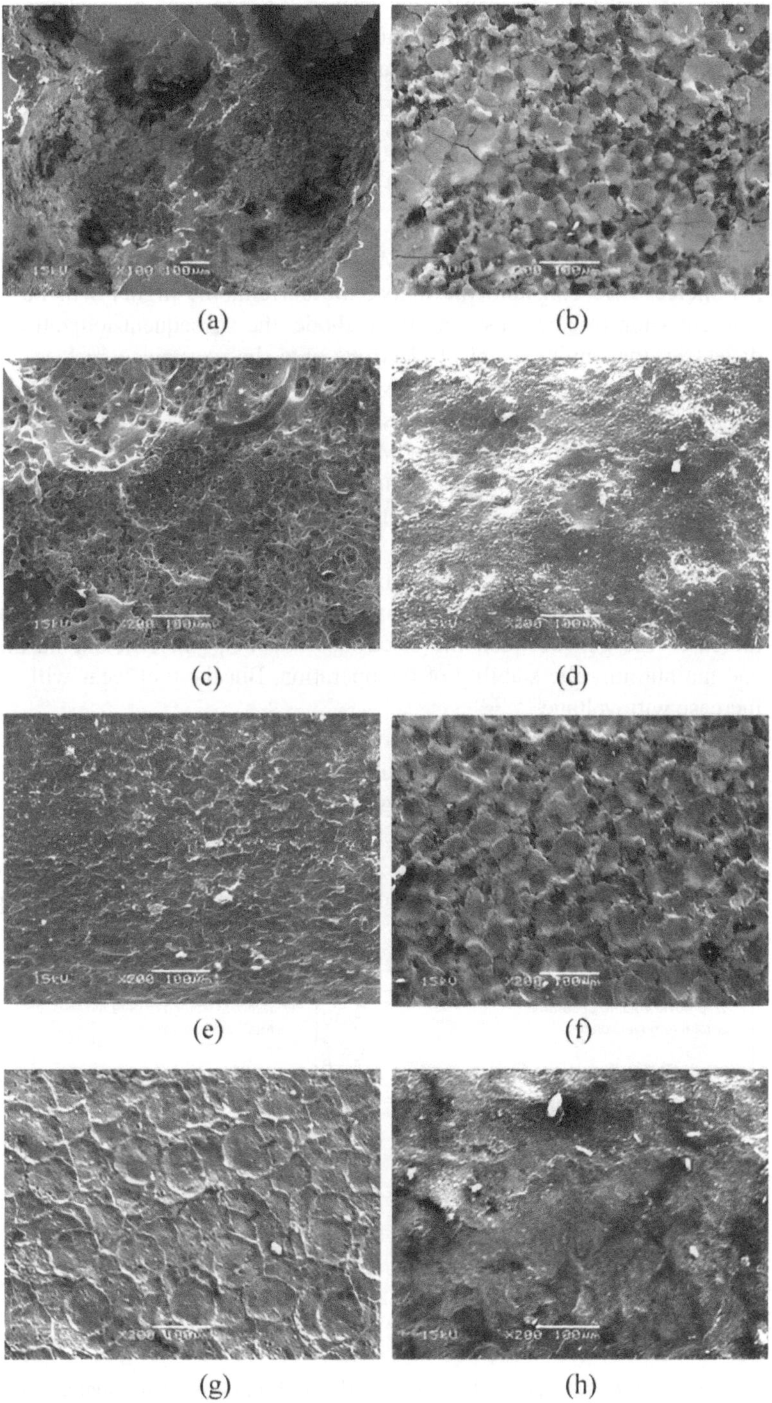

FIGURE 5.7 SEM images of AEM-EDM surface. (a) Al_2O_3, (b) ZrO_2, (c) TiO_2, (d) B_4C, (e) SiC, and (f) ZrO_2 (with Y_2O_3 addition)[22].

the material surface. This stress difference causes the formation of cracks and flaky material removal[23,24]. The thermal spalling mode of material removal in ceramics is not clearly understood because the formation of cracks in ceramics depends on a wide range of factors.

5.4.4 Effect of Process Parameters

a) Current

Current has a direct influence on MRR and surface finish. With all other parameters held constant, the higher the current, the higher will be the cold emission of electrons from the cathode, the subsequent ionization of dielectric, the impingement of electrons onto the workpiece surface, and hence the amount of material removed[20–22]. The material gets removed in the form of deeper and wider craters, so the surface finish of the workpiece reduces. The current in AEM-EDM of ceramics is in the range of 5–30 A[20,21]. Figure 5.8.b shows the influence of peak current on the material removal of Si_3N_4 ceramic and carbon conductive layer[25].

b) Voltage

Voltage has a direct influence on discharge energy, so an increase in voltage increases the amount of material removal from both electrodes. This increases the gap between the electrodes and helps in efficient flushing and maintaining the stability of the operation. But the tool wear will also increase with voltage[20–22,25].

c) Pulse on-time

Discharge energy also depends on the pulse on-time. The higher the pulse on-time, the higher will be the energy available to remove the material, but the material removal will be in the form of broader and deeper craters, so the surface roughness will be high[25]. Figure 5.8.a shows the influence of pulse on-time on MRR. Higher pulse on-time also means that the discharge

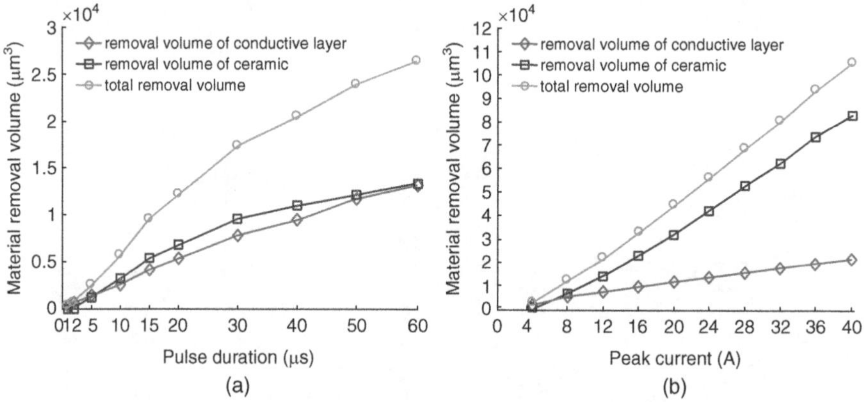

FIGURE 5.8 Effect of (a) pulse duration and (b) peak current on the material removal volume[25].

energy can penetrate the material for a longer time, so the HAZ and recast layer also increase.

d) Duty cycle
Increasing the duty cycle increases the material removal but reduces the pulse off-time, so flushing reduces, and the operation becomes erratic. So, a very high duty cycle reduces the MRR. The optimum duty cycle is between 40% and 60%.

e) Frequency
With all other parameters held constant and increasing frequency, the number of cycles increases and the discharge energy available is spread over all the cycles. So, with high frequencies, the material is removed in a controlled fashion with smaller craters, so the MRR is less but surface finish is high and vice versa.

5.4.5 Advantages and Disadvantages

EDM helps in achieving tight tolerances and accuracies, which was not the case with other machining processes. Complex and intricate shapes can be formed. The main advantage of using EDM on ceramics is the reduction in cost and time involved in machining rather than grinding.

The main disadvantage of EDM is that the sparks produced must be effectively controlled to avoid thermal and mechanical damage to the workpiece. Care should be taken to choose dielectrics that do not affect the ceramic material. Low MRR and high tool wear are very important drawbacks in EDM.

5.5 ELECTRIC DISCHARGE MACHINING OF COMPOSITES

Metal matrix composites (MMCs) and CMCs are widely machined using EDM[26]. PMCs are generally not used due to their low melting point and electrically insulative nature.

5.5.1 Electric Discharge Machining of Metal Matrix Composites (MMCs)

To overcome the difficulties in conventional machining of composites, unconventional machining processes came into existence. One of the well-established and commonly used advanced machining processes is EDM. MMCs being electrically conductive can be readily machined by EDM. The EDM of MMCs is similar to that of their parent metals and alloys, but the presence of reinforcement particles complicates and slows down the process. Copper, aluminum, iron, titanium, and magnesium are the commonly used matrix materials. Silicon carbide and alumina are the most widely used reinforcement materials[27].

SiC (silicon carbide) is the most common particle reinforcement used in aluminum-based MMCs. Extensive work has been done on the machining of SiC-aluminum composites[28–30]. Yan et al. investigated the effect of process parameters on MRR,

TWR, and surface roughness in EDM of Al_2O_3-reinforced Al6061 composites with a disk-like electrode[31]. A study on the EDM of WC-Co composites was carried out by Mahdavinejad et al.[32].

5.5.1.1 Material Removal Mechanism

The material removal mechanism followed in the EDM of MMCs is similar to that of monolithic metals and alloys. But the reinforcements have a shielding effect on the matrix. The very high electrothermal energy produced by the sparks melts the workpiece material and partially vaporizes it. The matrix material having a melting temperature less than that of the reinforcement melts readily, and due to this the reinforcement particles get loose and separated. These removed reinforcement particles then mix with the molten matrix material[33]. The liquid material comes in contact with the dielectric, cools, and deposits on the surface of the workpiece, forming a 'recast layer'. Usually, this layer has some entrapped debris, has higher strength, and affects the surface finish. It should be removed before the material can be used for any applications. The gaseous material removed from the workpiece condenses and forms spherical, hollow particles that are flushed away by the dielectric. This debris causes secondary discharges and makes the process erratic. These should be effectively flushed away for process stability, which is why flushing methods, flushing pressure, and pulse off-time become important.

5.5.1.2. Effect of Process Parameters

The primary process parameters in the EDM of MMCs are current, pulse on-time, and voltage. F Muller et al. studied the influence of current and pulse on-time on MRR and surface roughness. They conducted experiments on two grades of aluminum particle-reinforced metal matrix composites (PRMMCs) with different volume fractions of SiC (AA2618/SiC/20p and A356/SiC/35p) and unreinforced alloy (A356). Figure 5.9 shows that the MRR for PRMMCs increases with an increase in discharge current and pulse on-time up to a maximum value and then decreases.

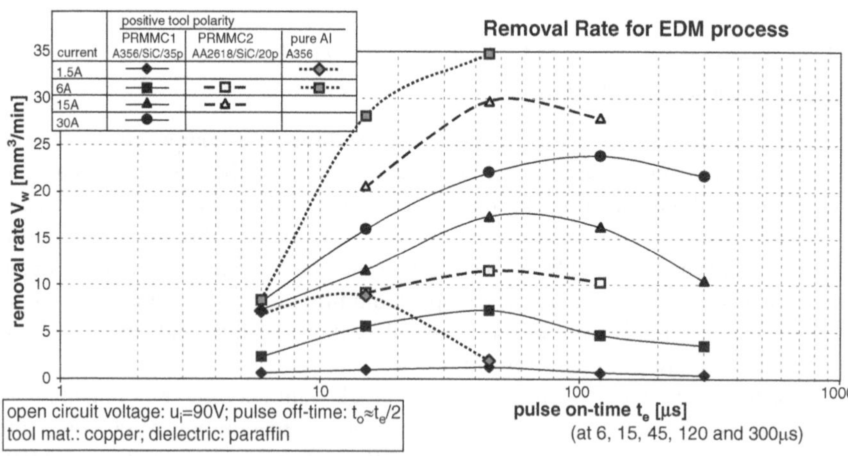

FIGURE 5.9 Effect of pulse on-time and current on the material removal rate[29].

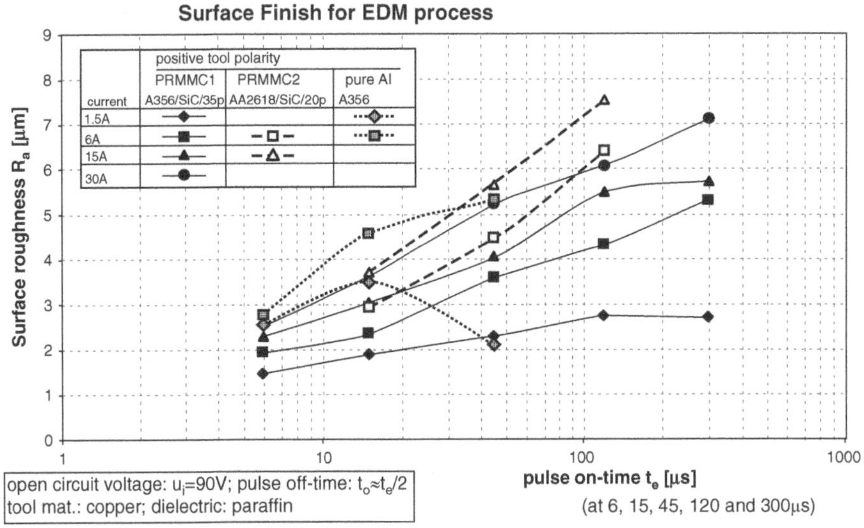

FIGURE 5.10 Effect of pulse on-time and current on the surface roughness[29].

The reason for this is that with higher current and longer pulse on-time, the matrix material (aluminum having a lower melting point) melts, and the reinforcement (SiC) particles are loosened. These particles due to insufficient flushing hinder the formation of subsequent sparks and affect the material removal[29].

Figure 5.10 shows the effect of pulse on-time and current on the surface roughness of the workpiece. As the current increases, the discharge energy increases and the thermal load acting on the electrodes increases. This causes material removal in the form of larger craters, which in turn increases the surface roughness. The increase in pulse on-time increases the removal per discharge in the form of large craters, which again causes increased surface roughness. The surface roughness of the unreinforced alloy is higher than that of the composites. This is because the total discharge energy is taken by the aluminum alloy and hence more material is removed in large craters. But in the case of composites, the reinforcement particles do not melt and due to their higher electrical resistivity shield the matrix, reducing the surface roughness[29].

The increase in the volume percentage of reinforcement in the matrix also reduces the MRR. The higher the volume percentage, the MRR will be lesser but the TWR and surface roughness will be higher. This is again due to the shielding effect of the reinforcement particles[31,34]. Figure 5.11 shows the influence of voltage on MRR. An increase in voltage increases the MRR. But with higher discharge currents and high voltage, the MRR reduces due to the accumulation of generated debris in the inter-electrode gap.

5.5.2 Electric Discharge Machining of Ceramic Matrix Composites (CMCs)

Advanced ceramics are basically ceramic-based composites where secondary materials are added to the parent ceramic to obtain enhanced properties. Ceramic composites

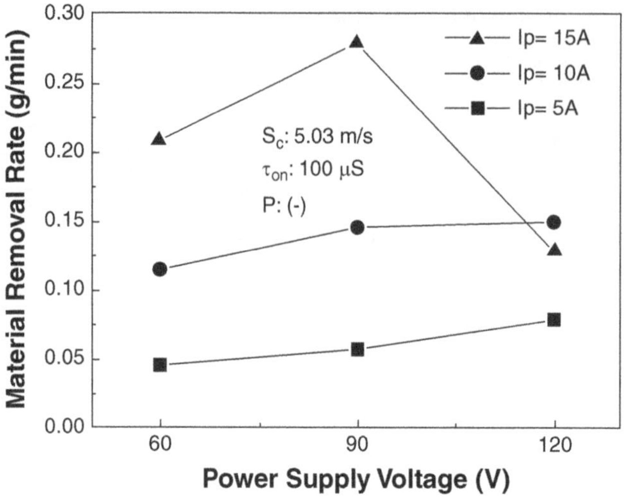

FIGURE 5.11 Effect of power supply voltage on MRR[31].

have better fracture toughness, ductility, thermal and electrical properties than traditional ceramics. But still they have very high hardness to be effectively machined by conventional methods. Unconventional machining of ceramic composites is a viable solution to overcome the limitations associated with conventional machining[35].

Ceramic composites containing electrically conductive reinforcements have better electrical conductivity than those containing electrically non-conductive reinforcements. Composites with higher electrical conductivity can be machined by normal EDM whereas insulative composites are machined by assisting electrode EDM. Carbon nanostructures like carbon nanotubes (CNTs), carbon nanoplatelets (CNPs), and graphene nanoplatelets are the commonly used reinforcement materials to increase electrical conductivity. Hanaoka et al. performed AEM and EDM processes on conventional Si_3N_4 ceramics and Si_3N_4 containing the above-said carbon nanostructures and compared the results. They studied the effect of electrical conductivities of the ceramics on the MRR, TWR, and surface roughness[36].

Andrea Gommeringer et al. studied the influence of tungsten carbide particle sizes and volume percentages on the characteristics of sinking and wire EDM in yttria-stabilized zirconia-tungsten carbide composites. Composites containing three different particle sizes of three different volume percentages were selected and subjected to EDM. Figure 5.12 shows the effect of particle size and volume percentage of tungsten carbide particles on MRR. The machinability by EDM increases with electrical conductivity, which is dependent on the volume percentage of the secondary particles. Particle sizes also have an influence on electrical conductivity, because the percolation capacity of smaller particles is higher than that of coarser particles. But very fine particles tend to agglomerate. Smaller-sized particles with higher volume percentage content show high electrical conductivity and hence good MRR[37].

Lauwers et al. investigated the different material removal mechanisms of Si_3N_4-based, ZrO_2-based, and Al_2O_3-based composites in deionized water and oil as

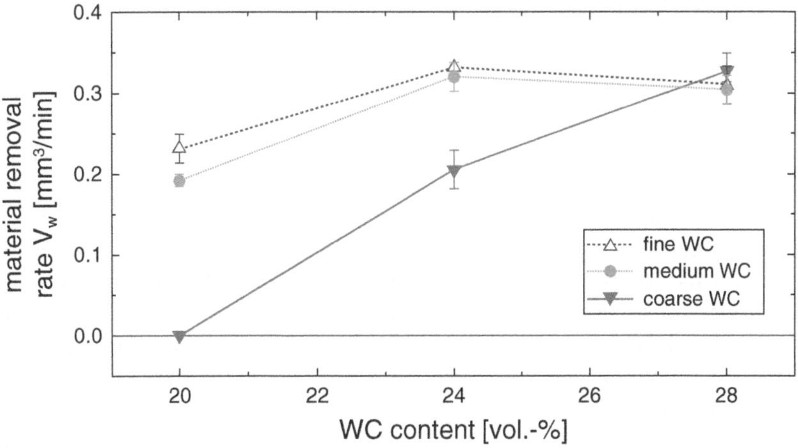

FIGURE 5.12 Effect of particle size and volume percentage on MRR[37].

dielectrics for different machining conditions. These composites contained TiCN and TiN as secondary conductive particles. They observed that in addition to the melting, evaporation, and thermal spalling mechanisms usually observed in ceramics, oxidation and decomposition were also observed in Si_3N_4-TiN composites. Oxidation reactions increased the MRR. The mechanism depended not only on the dielectric type and machining conditions but also on the discharge energies, pulse types, and pulse durations. Small discharge energies caused melting and evaporation whereas higher discharge energies caused thermal spalling. Melting and evaporation are characterized by craters, recast layer, and spherical debris while thermal spalling is characterized by cracks on the surface of the workpiece and irregular debris. Shorter pulse durations cause material removal by chemical reactions like oxidation/decomposition, and longer pulses cause melting[38].

5.5.3 Advantages and Disadvantages

The key advantage of using EDM is the ability to machine composites irrespective of their hardness. Since it is a non-contact type process, there are literally no forces acting on the workpieces. The common limitations in conventional machining of composites, such as fiber pull-out, debonding, delamination, particle cut-through, and particle pull-out, are effectively minimal in EDM. Complex and intricate features can be machined with EDM. MRR in EDM is very high compared to the conventional grinding of ceramic composites, and the cost involved is comparatively less. The primary limitation of EDM is that only electrically conductive materials can be machined, but that has been eliminated with the advent of AEM.

The problem of recast layer and HAZ is dominant in EDM. This can affect the mechanical and surface properties of the composites. The TWR is high in EDM, and it has to be effectively compensated to avoid form errors. Due to the high temperature involved, composites having less thermal conductivity undergo thermal shocks, which can damage them.

5.6 CONCLUSIONS

Though ceramics and composites have enhanced properties than their monolithic counterparts, their widespread usage is limited by their tedious and costly processing methods. The applicability of the EDM process for the machining of ceramics and composites has revolutionized the composites and ceramic industries. The capabilities and versatility of EDM processes have made them a basic and must-have machine tool in all processing industries. A significant amount of research has been and is being carried out on EDM processes, especially with respect to the machining of ceramics and composites. These materials are used in applications requiring a very good surface finish, and hence the effect of process parameters and reinforcements on surface finish is given importance. In addition to the conventional EDM processes, a number of advanced and hybrid processes like electro discharge milling[39], electro discharge drilling[40], electro-chemical discharge grinding[41], electro discharge diamond grinding[42], wire electro discharge grinding[43], and ultrasonic-assisted Electrochemical Discharge Machining[44] are also successfully adapted nowadays. There are still many unexplored avenues in the EDM of ceramics and composites, which can further the widespread use of these versatile materials.

REFERENCES

[1] Callister, W. D.; Rethwisch, D. G. *Materials Science and Engineering*; Wiley, 2014.

[2] Chawla, K. K. *Composite Materials: Science and Engineering, Third Edition*; Springer, 2012. https://doi.org/10.1007/978-0-387-74365-3.

[3] Agarwal, S.; Rao, P. V. Experimental Investigation of Surface/Subsurface Damage Formation and Material Removal Mechanisms in SiC Grinding. *International Journal of Machine Tools and Manufacture*, 2008, *48* (6), 698–710. https://doi.org/10.1016/j.ijmachtools.2007.10.013.

[4] Malkin, S.; Hwang, T. W. Grinding Mechanisms for Ceramics. *Annals of the CIRP*, 1996, *45* (2), 569–580.

[5] Tvergaard, V.; Xia, Z. C.; Hutchinson, J. W. Cracking Due to Localized Hot Shock. *Journal of the American Ceramic Society*, 1993, *76* (3), 729–736. https://doi.org/10.1111/j.1151-2916.1993.tb03667.x.

[6] Dubey, A. K.; Yadava, V. Laser Beam Machining-A Review. *International Journal of Machine Tools and Manufacture*, 2008, *48* (6), 609–628. https://doi.org/10.1016/j.ijmachtools.2007.10.017.

[7] Tsai, C. H.; Chen, H. W. Laser Milling of Cavity in Ceramic Substrate by Fracture-Machining Element Technique. *Journal of Materials Processing Technology*, 2003, *136* (1–3), 158–165. https://doi.org/10.1016/S0924-0136(03)00133-X.

[8] Thoe, T. B.; Aspinwall, D. K.; Wise, M. L. H. Review on Ultrasonic Machining. *International Journal of Machine Tools and Manufacture*, 1998, *38* (4), 239–255. https://doi.org/10.1016/S0890-6955(97)00036-9.

[9] Chen, L.; Siores, E.; Wong, W. C. K. Kerf Characteristics in Abrasive Waterjet Cutting of Ceramic Materials. *International Journal of Machine Tools and Manufacture*, 1996, *36* (11), 1201–1206. https://doi.org/10.1016/0890-6955(95)00108-5.

[10] Zeng, J.; Kim, T. J. An Erosion Model for Abrasive Waterjet Milling of Polycrystalline Ceramics. *Wear*, 1996, *199* (2), 275–282. https://doi.org/10.1016/0043-1648(96)07223-7.

[11] Nicolais, L.; Sheikh-Ahmad, J. Y.; Davim, J. P. Cutting and Machining of Polymer Composites. In *Wiley Encyclopedia of Composites*, Nicolais, L. (Ed.). 2012 July. https://doi.org/10.1002/9781118097298.weoc061.

[12] Wang, J. Abrasive Waterjet Machining of Polymer Matrix Composites-Cutting Performance, Erosive Process and Predictive Models. *The International Journal of Advanced Manufacturing Technology,* 1999, *15,* 757–768.
[13] Oyelola, O.; Crawforth, P.; M'Saoubi, R.; Clare, A. T. Machining of Functionally Graded Ti6Al4V/ WC Produced by Directed Energy Deposition. *Additive Manufacturing,* 2018, *24,* 20–29. https://doi.org/10.1016/j.addma.2018.09.007.
[14] Nicholls, C. J.; Boswell, B.; Davies, I. J.; Islam, M. N. Review of Machining Metal Matrix Composites. *International Journal of Advanced Manufacturing Technology,* 2017, *90* (9–12), 2429–2441. https://doi.org/10.1007/s00170-016-9558-4.
[15] Jain, V. K. *Advanced Machining Processes*; Allied Publishers, 2009.
[16] Guitrau, E. B. *The EDM Handbook*; Carl Hanser GmbH, 2009.
[17] Liu, K.; Lauwers, B.; Reynaerts, D. Process Capabilities of Micro-EDM and Its Applications. *The International Journal of Advanced Manufacturing Technology,* 2010, *47* (1–4), 11–19.
[18] Grigoriev, S. N.; Hamdy, K.; Volosova, M. A.; Okunkova, A. A.; Fedorov, S. V. Electrical Discharge Machining of Oxide and Nitride Ceramics: A Review. *Materials & Design,* 2021, *209,* 109965. https://doi.org/https://doi.org/10.1016/j.matdes.2021.109965.
[19] Moulson, A. J.; Herbert, J. M. *Electroceramics: Materials, Properties, Applications,* 2nd ed.; John Wiley & Sons, 2003.
[20] Hösel, T.; Müller, C.; Reinecke, H. Spark Erosive Structuring of Electrically Nonconductive Zirconia with an Assisting Electrode. *CIRP Journal of Manufacturing Science and Technology,* 2011, *4* (4), 357–361. https://doi.org/10.1016/j.cirpj.2011.05.005.
[21] Mohri, N.; Fukuzawa, Y.; Tani, T.; Saito, N.; Furutani, K. Assisting Electrode Method for Machining Insulating Ceramics. *CIRP Annals - Manufacturing Technology,* 1996, *45* (1), 201–204. https://doi.org/10.1016/S0007-8506(07)63047-9.
[22] Kucukturk, G.; Cogun, C. A New Method for Machining of Electrically Nonconductive Workpieces Using Electric Discharge Machining Technique. *Machining Science and Technology,* 2010, *14* (2), 189–207. https://doi.org/10.1080/10910344.2010.500497.
[23] Petrofes, N. F. *Shaping Advanced Ceramics with Electrical Discharge Machining*; Texas A&M University, 1989.
[24] Gadalla, A. M. Thermal Spalling during Electro-Discharge Machining of Advanced Ceramics and Ceramic-Ceramic Composites. *Machining of Composite Materials Symposium*, Chicago, 1992.
[25] Hou, P. J.; Guo, Y. F.; Sun, L. X.; Deng, G. Q. Simulation of Temperature and Thermal Stress Filed during Reciprocating Traveling WEDM of Insulating Ceramics. *Procedia CIRP,* 2013, *6,* 410–415. https://doi.org/https://doi.org/10.1016/j.procir.2013.03.010.
[26] König, W.; Dauw, D. F.; Levy, G.; Panten, U. EDM-Future Steps towards the Machining of Ceramics. *CIRP Annals - Manufacturing Technology,* 1988, *37* (2), 623–631. https://doi.org/10.1016/S0007-8506(07)60759-8.
[27] VG, S. P. Effect of Ultrasonic Treatment during Stir Casting on Mechanical Properties of AA6063-SiC Composites. *Materials Chemistry and Physics,* 2023, *294,* 126977.
[28] Shanmuga Priyan V. G.; Kanmani Subbu S. Microscopic Investigation on Micro Electric Discharge Milling of AA6063-SiC Composites. *Machining Science and Technology,* 2024, *28* (4), 656–681. https://doi.org/10.1080/10910344.2024.2369861
[29] Müller, F.; Monaghan, J. Non-Conventional Machining of Particle Reinforced Metal Matrix Composite. *Journal of Materials Processing Technology,* 2001, *118* (1–3), 278–285. https://doi.org/10.1016/S0924-0136(01)00941-4
[30] Hocheng, H.; Lei, W. T.; Hsu, H. S. Preliminary Study of Material Removal in Electrical-Discharge Machining of SiC/Al. *Journal of Materials Processing Technology,* 1997, *63* (1–3), 813–818. https://doi.org/10.1016/S0924-0136(96)02730-6.
[31] Yan, B. H., Wang, C. C., Liu, W. D., & Huang, F. Y. (2000). Machining Characteristics of Al_2O_3/6061Al Composite Using Rotary EDM with a Disklike Electrode. *The*

International Journal of Advanced Manufacturing Technology, 2000, *16*, 322–333. https://doi.org/10.1007/s001700050164.
[32] Mahdavinejad, R. A.; Mahdavinejad, A. ED Machining of WC-Co. *Journal of Materials Processing Technology*, 2005, *162–163* (SPEC. ISS.), 637–643. https://doi.org/10.1016/j.jmatprotec.2005.02.211.
[33] Shanmuga Priyan V.G.; S. Kanmani Subbu. Effect of Tool Feed Rate and Tool Rotation Speed on the Microchannel Geometry of AA6063-SiC Composites Machined by Micro ED Milling. *Advances in Materials and Processing Technologies*, 2024, *11* (2), 724–750. https://doi.org/10.1080/2374068X.2024.2321701.
[34] Garg, R. K.; Singh, K. K.; Sachdeva, A.; Sharma, V. S.; Ojha, K.; Singh, S. Review of Research Work in Sinking EDM and WEDM on Metal Matrix Composite Materials. *International Journal of Advanced Manufacturing Technology*, 2010, *50* (5–8), 611–624. https://doi.org/10.1007/s00170-010-2534-5.
[35] Ming, W.; Jia, H.; Zhang, H.; Zhang, Z.; Liu, K.; Du, J.; Shen, F.; Zhang, G. A Comprehensive Review of Electric Discharge Machining of Advanced Ceramics. *Ceramics International*, 2020, *46* (14), 21813–21838. https://doi.org/https://doi.org/10.1016/j.ceramint.2020.05.207.
[36] Hanaoka, D.; Fukuzawa, Y.; Ramirez, C.; Miranzo, P.; Osendi, M. I.; Belmonte, M. Electrical Discharge Machining of Ceramic/Carbon Nanostructure Composites. *Procedia CIRP*, 2013, *6*, 95–100. https://doi.org/https://doi.org/10.1016/j.procir.2013.03.033.
[37] Gommeringer, A.; Schmitt-Radloff, U.; Ninz, P.; Kern, F.; Klocke, F.; Schneider, S.; Holsten, M.; Klink, A. ED-Machinable Ceramics with Oxide Matrix: Influence of Particle Size and Volume Fraction of the Electrical Conductive Phase on the Mechanical and Electrical Properties and the EDM Characteristics. *Procedia CIRP*, 2018, *68*, 22–27. https://doi.org/https://doi.org/10.1016/j.procir.2017.12.016.
[38] Lauwers, B.; Kruth, J. P.; Liu, W.; Eeraerts, W.; Schacht, B.; Bleys, P. Investigation of Material Removal Mechanisms in EDM of Composite Ceramic Materials. *Journal of Materials Processing Technology*, 2004, *149* (1–3), 347–352. https://doi.org/10.1016/j.jmatprotec.2004.02.013.
[39] D'Urso, G.; Giardini, C.; Lorenzi, S.; Quarto, M.; Sciti, D.; Silvestroni, L. Micro-EDM Milling of Zirconium Carbide Ceramics. *Precision Engineering*, 2020, *65*, 156–163. https://doi.org/10.1016/j.precisioneng.2020.06.002.
[40] Malayath, G.; Katta, S.; Sidpara, A. M.; Deb, S. Length-Wise Tool Wear Compensation for Micro Electric Discharge Drilling of Blind Holes. *Measurement (Lond)*, 2019, *134*, 888–896. https://doi.org/10.1016/j.measurement.2018.12.047.
[41] Puri, A. B.; Banerjee, S. Multiple-Response Optimisation of Electrochemical Grinding Characteristics through Response Surface Methodology. *The International Journal of Advanced Manufacturing Technology*, 2013, *64* (5), 715–725. https://doi.org/10.1007/s00170-012-4065-8.
[42] Yadav, R. S.; Yadava, V. Experimental Investigations on Electrical Discharge Diamond Peripheral Surface Grinding (EDDPSG) of Hybrid Metal Matrix Composite. *Journal of Manufacturing Processes*, 2017, *27*, 241–251. https://doi.org/https://doi.org/10.1016/j.jmapro.2017.04.004.
[43] Luo, H.; Zou, H.; Liu, B.; Yue, X. Fabrication of Micro Rotary Structures by Wire Electrochemical Grinding. *Electrochemistry communications*, 2020, *116*, 106745. https://doi.org/https://doi.org/10.1016/j.elecom.2020.106745.
[44] Hung, J.-C.; Zhang, Y.-W. Ultrasonic-Assisted Electrochemical Discharge Grinding and Broaching for Machining Quartz Square Microholes. *Journal of Materials Research and Technology*, 2023, *25*, 1782–1799. https://doi.org/https://doi.org/10.1016/j.jmrt.2023.06.027.

6 Multi-Objective Optimization of Wire-EDM Using ANN-NSGA-II Hybrid Approach

Abhijit Saha, Dipika Sarkar, and Prosun Mandal

6.1 INTRODUCTION

Hardfacing, also known as hard surfacing, is the process of applying resistance to a wear surface over a component susceptible to abrasion by depositing a hardfacing substance on the base metal. In the previous few decades, research on improving the properties of hardfacing materials has greatly advanced to address these challenges. The majority of the necessary qualities, such as strong chemical or heat stabilities, high abrasion resistance, and superior mechanical properties, are all concurrently maintained at a high degree, which is a crucial attribute of efficient hardfacing. The choice of the coating materials and their chemical bonding for a certain application determines the effectiveness of the hardfacing procedure. To increase the resistance to corrosion and wear of the components, several welding processes, including oxyacetylene welding, submerged arc welding, plasma arc welding, and manual metal arc welding, have been widely applied in large-scale manufacturing. Hardfacing materials are widely used to protect screw conveyers, sinter plant fans, sinter crusher bars, furnace chutes, mixer blades, exhaust fans, and grizzly bars. Because of the nano-composite, the mechanical characteristics of hardfacing material of fine borides, carbides, and boron-carbides are mechanically enhanced (greater hardness, better toughness, and amazing wear characteristics even at high temperatures). It is exceedingly challenging to cut complicated and compound forms in hard materials using standard methods because of the high wear resistance and hardness of hardfacing material. Wire-EDM (wire electrical discharge machining) process is one of the non-traditional machining techniques, useful in the electro-thermal method of cutting such challenging electrically conductive materials. The WEDM technique has faced challenges in recent years due to the introduction of newer, more exotic materials in an industrial environment.

It can be challenging to create a precise method and apply it to choose the best process parameters for the WEDM process because of the non-linear and complex

DOI: 10.1201/9781003769675-6

stochastic behavior of output responses and input process parameters. To determine the best parameter choices and the relationship between the response parameters and the input process parameters, many prediction models are being applied. So, researchers concentrated on predicting reactions using mathematical models like response surface methodology (RSM). The RSM was also used to develop a mathematical model for the wire-EDM process by different researchers [1–4]. Devarasiddappa and Chandrasekaran [5] optimized the output responses of wire-cut EDM using principal component analysis (PCA) coupled withthe grey relational analysis (GRA) technique. Vijay Praveen et al. [6] also used the same PCA coupled with GRA techniques to optimize the wire-EDM on Ni-coated Al_2O_3p/AA7075 MMCs. There are various assumptions that need to be considered in statistical methods for the development of empirical and regression models. These processes also have numerous drawbacks. Sometimes, higher-order curve fitting needs to be done with non-linear data and transformation of the data is required, which is a problematic task. Along with that, noisy data cannot be best fitted in a regression analysis model [7]. Due to that, modeling with highly intelligent techniques like artificial neural networks (ANNs) has become very popular in recent years. A number of researchers used ANN to develop the model of different manufacturing processes [8,9]. The extremely non-linear process parameters lead the researcher to adopt meta-heuristic optimization techniques. Genetic algorithm (GA) is one of the popular techniques widely used for optimizing the processes. Wang et al. [10] developed and implemented a mechanism for combining the important capabilities of ANN and GA to model and optimize the EDM process. A system integrating the key functions of ANNs and GAs has been created and applied. Assarzadeh and Ghoreishi [11] utilized a 3-6-4-2 backpropagation (BP) neural network architecture combined with GA to predict and optimize material removal rate (MRR) and surface roughness (SR) in the EDM of BD3 steel. Their findings revealed that the ANN-GA model produced highly accurate predictions for optimized MRR and SR. Gurram et al. [12] designed an ANN model for the EDM machining of materials like Ti-6Al-4V, HE-15, 15CDV6, and M-25 to forecast SR. The absolute error of the ANN model, when paired with GA, was less than 2%, compared to over 5% with the standalone ANN model. Similarly, Somashekhar et al. [13] developed an ANN model to predict MRR, which was then optimized using GA. The 4-6-6-1 ANN architecture was found to be the best model, and this model was employed for converging to a global optimal solution. Several researchers also investigated and optimized the machining parameters in various machining processes [14–17]. The optimal machining conditions for nanostructured hardfaced material with the help of wire-EDM have an important role in improving the performance. Literature shows that a very efficient prediction model can be developed using the ANN technique. GA is a meta-heuristic optimization technique widely used to find optimal machining parameters. In this paper, both ANN and multi-objective genetic algorithm (NSGA II) were used to develop a prediction model and solve multi-objective optimization problem, respectively.

6.2 ANN-NSGA-II HYBRID APPROACH

The wire-EDM process is inherently random and unpredictable, making it challenging to establish clear input-output relationships using conventional methods.

In contrast, the ANN approach is an advanced, flexible technique capable of modeling complex, non-linear relationships between input parameters and output results. Inspired by the human central nervous system, ANNs can learn machine behavior and recognize patterns [18]. ANN architectures consist of various layers (input, hidden, and output), each containing multiple neurons connected by weights. Training a neural network involves adjusting these weights to facilitate learning and pattern recognition [19]. Several algorithms are available for neural network training. In this study, the neural network is trained using the Levenberg-Marquardt (LM) BP algorithm. The output of the neural network is computed using the following formula:

$$O_j = f\sum_{i=1}^{n} w_{ij}x_i + b_i \tag{6.1}$$

where O_j represents the output of the jth neuron in any layer, f stands for the activation functions, such as linear, tansigmoidal, or logsigmoidal, n is the input to the jth neuron, w_{ij} is the corresponding weight, x_i is the output from the ith neuron, and b_i is the associated bias.

The neural network's performance is often assessed using the mean square error (MSE), calculated as follows:

$$\text{MSE} = \sum_{1}^{n}\left\{y_i - f(x_i)\right\}^2 \tag{6.2}$$

where y_i is the experimental or target output and $f(x_i)$is the predicted output from the network. A neural network model is selected based on its minimal MSE and high correlation value (R-squared). Selecting the appropriate number of neurons in the hidden layer is crucial for minimizing MSE, and this is typically determined through trial and error to achieve the best neural network performance. In this study, input parameters for the WEDM process included discharge pulse time (T_{on}), discharge stop time (T_{off}), servo voltage (SV), wire tension (Wt), and wire feed rate (Wf), with corresponding outputs being MRR, machining time, and SR. The experimental data, obtained using an L25 Taguchi orthogonal array design, was used to train a neural network. The best network architecture was determined through a trial-and-error approach, with the selection based on minimizing the MSE. The ANN model was implemented in MATLAB.

Genetic algorithm (GA), an evolutionary computation technique, is capable of solving both linear and non-linear stochastic problems by simulating the concept of natural selection, specifically Darwin's "survival of the fittest." GA can tackle both single- and multi-objective optimization problems under various constraints [20–22]. The process evolves a population of potential solutions over multiple iterations. It begins with a randomly generated set of solutions (chromosomes), each representing different process parameters. The fitness of each chromosome is assessed through objective functions and through crossover and mutation operations, and a new population (offspring) is generated. In the crossover step, two chromosomes are split and recombined to produce new solutions. Mutation is the process of randomly changing

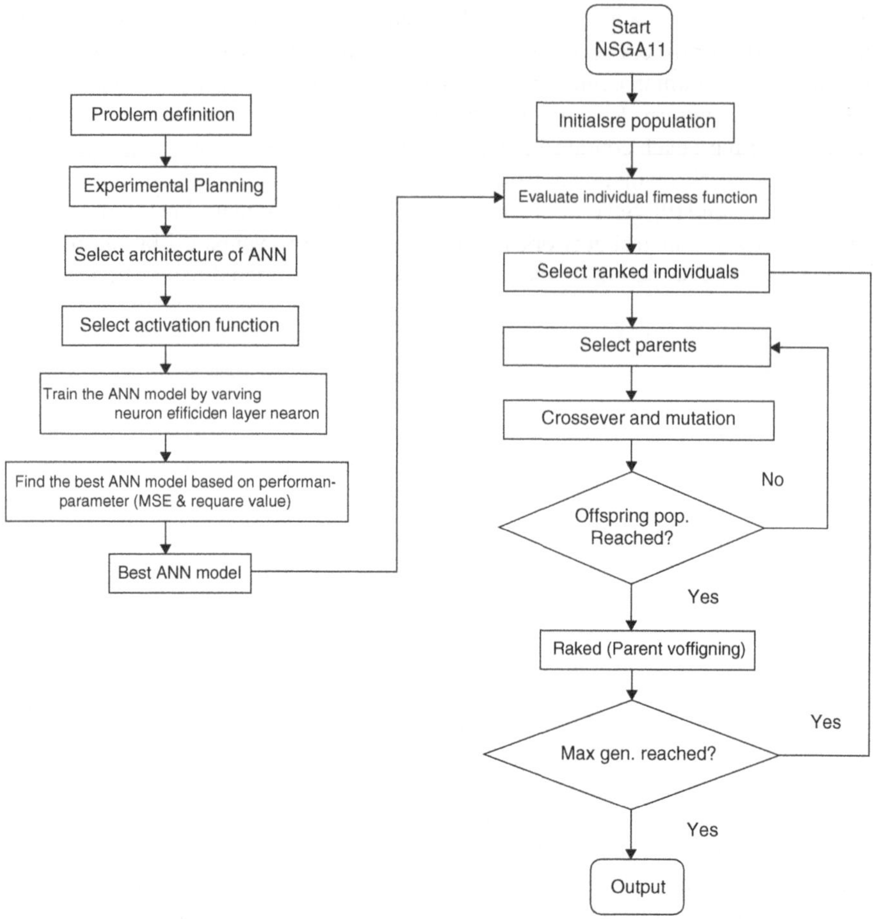

FIGURE 6.1 Flowchart for ANN-NSGA-II hybrid approach.

genes in an individual chromosome to maintain diversity. In this way, the population of individual solutions is repeatedly modified in successive generations. This process is continued till the reaching of global optimal solutions or obtaining the stopping criteria [23]. The flowchart (Figure 6.1) shows the ANN-NSGA-II hybrid approach for modeling and then solving the multi-objective optimization problem of the wire-EDM process. To demonstrate and validate the effectiveness of the ANN-NSGA-II hybrid method, the authors have considered the practical example of the WEDM process from the literature [1].

6.3 RESULT AND DISCUSSION

6.3.1 Modeling with ANN

The ANN is the most widely used machine learning technique for the prediction of output responses for different combinations of input process parameters [24–25].

Different ANN parameters, namely the hidden layer, number of neurons in the hidden layer, training algorithm, have a significant influence on obtaining the best model. MATLAB R2015a is utilized to build and train the ANN model. BP LM (trainlm) is used for training purposes. The structure of the ANN and the quantity of neurons in the hidden layer are determined through multiple trial-and-error attempts. The minimum MSE and a higher value of correlation coefficient (R) are the selection criteria for the ANN architecture [26–37]. Table 6.1 provides an overview of various ANN architectures along with their corresponding MSE for training and correlation coefficient (R). The optimal values, with the lowest MSE of 0.0000564433 and the highest R value of 0.999999, were achieved using the 5-14-3-3 ANN configuration. This suggests that the 5-14-3-3 architecture is the most effective model. Figure 6. 2 illustrates the 5-14-3-3 ANN structure, while Figure 6.3 shows the regression plot for the model's training phase. The plot demonstrates the strong correlation between the network's outputs and experimental data, with an R value of 1 for the 5-14-3-3 model, indicating excellent agreement between predicted and actual results. Consequently, the 5-14-3-3 ANN architecture is highly capable of accurately predicting MRR, machining time, and SR in the wire-EDM process.

TABLE 6.1
MSE and R-square value for different architectures

ANN architecture	MSE-training	*R*-square value
5-10-3-3	0.927426	0.99849
5-11-3-3	0.653211	0.998799
5-12-3-3	0.019226	0.999967
5-13-3-3	0.334559	0.999512
5-14-3-3	**0.0000564433**	**0.999999**
5-15-3-3	0.423088	0.999557
5-16-3-3	0.054174	0.999902
5-17-3-3	0.380668	0.999932
5-18-3-3	0.0314058	0.999894
5-19-3-3	0.0564274	0.999907
5-20-3-3	0.20306	0.999669

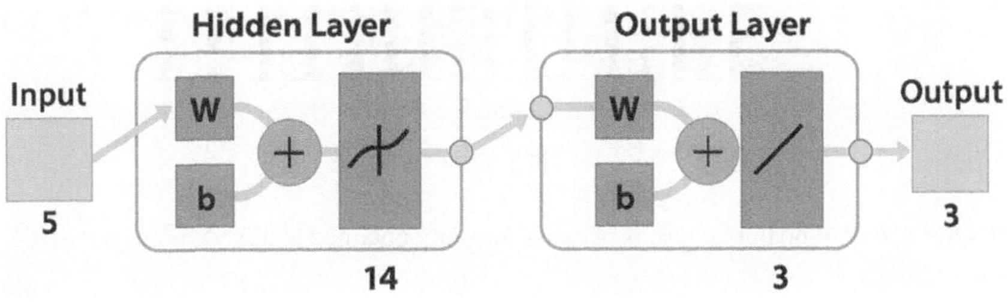

FIGURE 6.2 ANN architecture for the prediction of output responses.

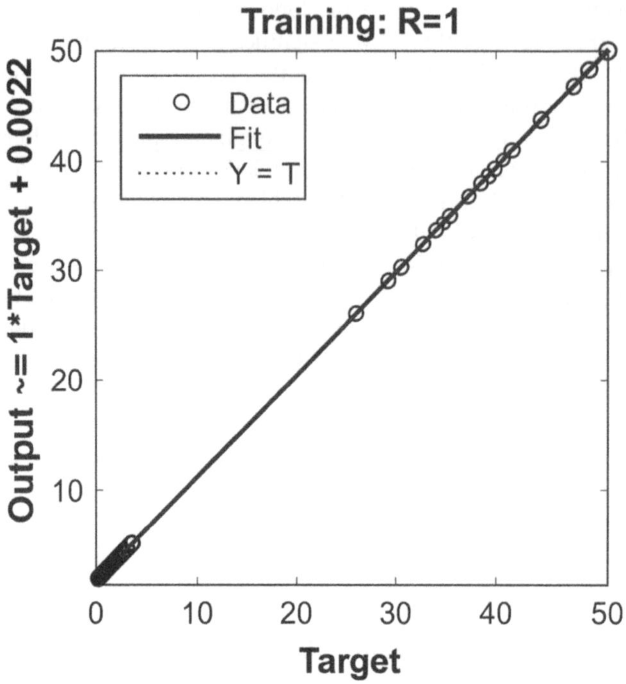

FIGURE 6.3 Regression plot for 5-14-3-3 ANN model.

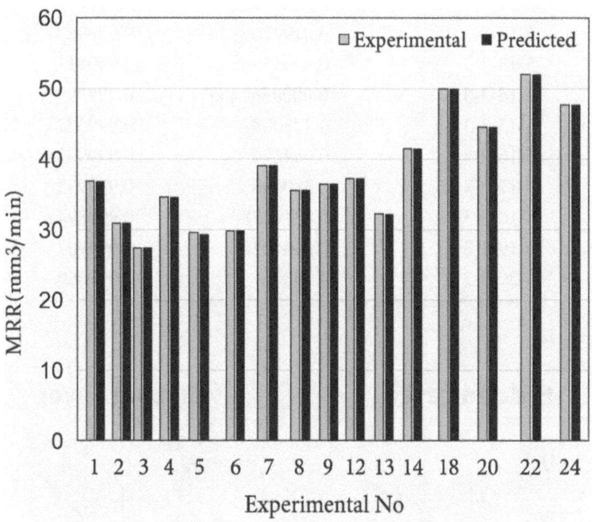

FIGURE 6.4 Comparative results of experimentally obtained and ANN predicted MRR.

A comparison between the MRR values obtained experimentally and those predicted by the ANN model (3-12-2-2) is illustrated in the bar chart (Figure 6.4). This comparison includes all experimental data. In the chart, the X-axis represents the experiment numbers, while the Y-axis shows the MRR values in mm³/min. It can be seen that the predicted MRR values from the ANN model closely match the experimental results for most experiments. Similar comparisons are made for machining time (min) and SR (μm) in Figures 6.5 and 6.6, respectively. Based on these findings,

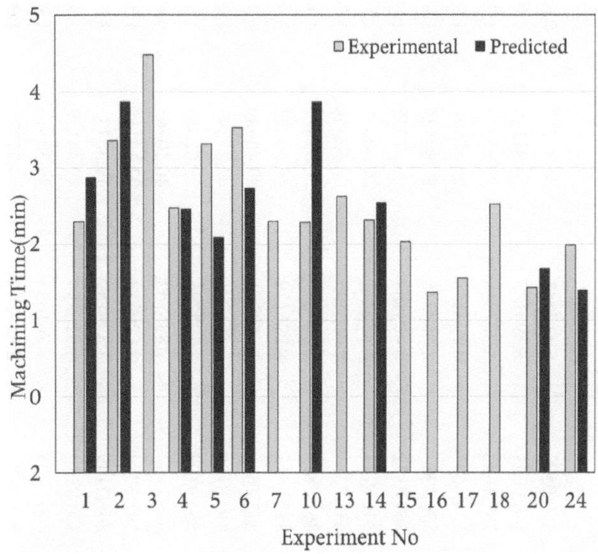

FIGURE 6.5 Comparative results of experimentally obtained and ANN predicted machining time.

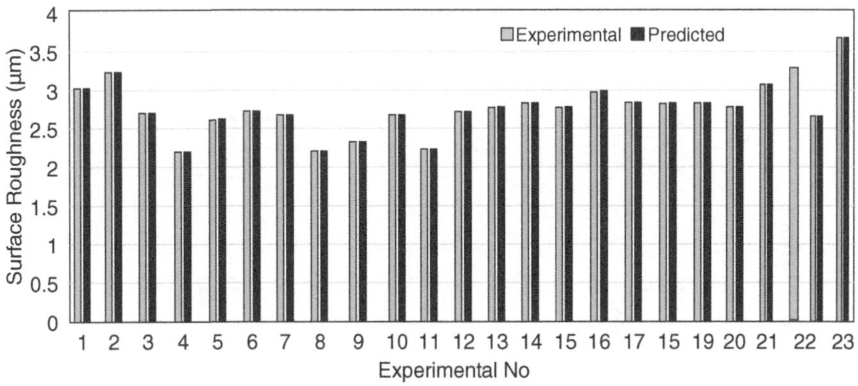

FIGURE 6.6 Comparative results of experimentally obtained and ANN predicted surface roughness.

TABLE 6.2
Pareto-optimal solutions

Sl. No	T_{on}	T_{off}	SV	Wt	Wf	MRR	Machining time	Surface roughness
1	3.162245	8.09868	38.04214	860.0507	8.517042	36.82268	0.688324257	2.495840238
2	2.895818	11.11227	35.05239	548.9998	8.393183	24.91119	1.969161507	0.557494783
3	3.468567	8.663942	37.27067	789.1683	8.367818	35.55324	1.363316183	2.255059305
4	2.799037	9.694169	35.20658	634.3647	8.724148	25.57504	1.916433826	1.385366266
5	3.073378	9.214437	35.87851	559.296	8.868346	27.64525	1.699540315	1.686737425
6	3.126386	9.704732	35.4514	567.8372	8.403864	26.87525	1.868485417	1.009329155
7	1.873049	10.74697	40.64649	846.8111	8.925221	20.60096	2.703550668	2.825495405
8	3.198079	9.289483	35.44347	633.23	8.466145	28.60658	1.761430057	1.564675214
9	1.174863	11.69367	43.58791	861.5879	8.567649	20.97317	4.15715932	2.682580273
10	1.647363	11.928	40.77731	833.1413	8.996179	17.76223	3.413310691	3.311166538
11	3.330914	8.188109	38.51314	862.1679	7.967301	37.62989	0.81971518	2.090937819
12	4.020377	8.165143	38.05021	860.8175	7.561275	40.87895	1.056495922	1.803095259
13	3.231669	9.061374	35.77326	553.9107	8.824735	29.22069	1.639214967	1.785255721
14	4.12236	8.147512	38.8978	866.253	7.835752	40.10663	0.921140444	1.964178174
15	3.201722	8.904614	36.42698	692.8924	8.620318	32.14563	1.545709928	2.368413465
16	1.307058	11.66687	43.2092	866.2149	8.879159	19.8073	3.787714858	2.924309217
17	1.880643	10.99698	38.38147	870.033	8.928954	19.44159	2.90638264	3.350726572
18	3.91664	8.309004	36.53013	860.7795	7.761655	39.38316	1.131895887	1.877039614
19	2.800708	10.66501	35.10851	558.448	8.659244	24.21376	1.999680485	0.663978547
20	2.544927	9.36447	38.62474	834.9261	8.932434	31.00377	1.445317039	2.945757345
21	2.340679	9.264133	38.42521	861.5403	8.8798	29.76462	1.574106051	3.017644128
22	1.873948	10.83594	39.20045	853.2413	8.927032	19.76454	2.792162078	3.079808558
23	2.50102	11.73312	37.65175	561.3052	8.953038	21.47194	2.446427599	0.882779281

(Continued)

TABLE 6.2
(Continued)

Sl. No	T_{on}	T_{off}	SV	Wt	Wf	MRR	Machining time	Surface roughness
24	1.851284	11.78323	40.64707	836.4488	8.985444	18.14037	3.218186951	3.234664827
25	3.089003	9.339437	35.92538	559.421	8.962096	27.30918	1.794364571	1.674246758
26	2.399291	11.76231	37.28826	556.2749	8.80798	21.91271	2.361002145	0.830312422
27	3.444852	9.174186	35.26701	552.2394	8.405272	30.58585	1.608854298	1.373775147
28	2.567796	10.30735	37.1297	563.0319	8.841182	23.54645	2.093949269	1.218883303
29	1.976284	11.67386	40.64707	836.3238	8.954194	18.66476	3.112026561	3.144739042
30	1.178225	11.64121	41.8996	830.8602	8.605256	20.51565	4.023880319	2.900497462
31	2.577048	10.82197	36.53918	592.1009	8.825007	23.13011	2.173697528	0.971662228
32	1.844655	11.10204	39.20111	856.1779	8.917239	19.13325	2.944506385	3.191484297
33	4.020377	8.165143	38.05021	860.8175	7.311275	42.5368	1.253207721	1.796403035
34	1.873948	11.08594	39.26295	853.3663	8.677032	19.85626	3.040390251	2.9389966
35	2.123948	10.83594	39.20045	853.2413	8.989532	20.6548	2.641175572	3.139385032

the 5-14-3-3 ANN model is considered effective in predicting MRR, machining time, and SR, making it suitable for use as a fitness function in NSGA-II to optimize solutions.

6.3.2 Optimization Process Using ANN-NSGA-II

In the current study, MRR, machining time, and SR are the output variables that need to be optimized based on various process parameters, including T_{on}, T_{off}, SV, Wt, and Wf. To achieve the best output, the correct selection of these input parameters is crucial. The NSGA-II technique is used to determine the optimal process parameters within the defined boundaries. Previously, an ANN model was developed to predict MRR, machining time, and SR in the wire-EDM process. This ANN model is now applied as a fitness function in the GA toolbox of MATLAB R2015a to find the optimal process parameters. The initial population size is set to 100 with a double vector population type. The optimal parameters are derived within specified upper and lower limits, which are as follows:

$8 \le T_{off} \le 12$

$0.3 \le T_{on} \le 5$

$35 \le SV \le 44$

$500 \le Wt \le 900$

$5 \le Wf \le 9$

A set of optimal solutions is obtained using these input parameters and boundary conditions. Table 6.2 displays the corresponding Pareto-optimal solutions.

6.4 CONCLUSION

In this study, the NSGA-II algorithm integrated with an ANN was effectively applied to determine the optimal process parameters for wire-EDM. Initially, an ANN model was developed, and the most accurate model was chosen to serve as the fitness function for NSGA-II. Key input parameters for wire-EDM, including T_{on}, T_{off}, SV, Wt, and Wf, were considered, while MRR, machining time, and SR were selected as output parameters. The ANN model was trained using experimental data, and the optimal model was identified through trial and error. The 5-14-3-3 ANN model was found to be the best-performing model and was used in NSGA-II as the fitness function. By combining this ANN model with NSGA-II, a set of optimal process parameters was determined, enabling operators to choose the best settings for wire-EDM to achieve desired results.

REFERENCES

[1] A. Saha, and S. C. Mondal, "Multi-objective optimization in WEDM process of nano-structured hardfacing materials through hybrid techniques," *Measurement*, 94 46–59 (2016).

[2] A. Kumar, T. Soota, and J. Kumar, "Optimisation of wire-cut EDM process parameter by Grey-based response surface methodology," *International Journal of Industrial Engineering,* 14(4) 821–829 (2018).

[3] D. Ganesh, S. Zaware, U. Dabade, and S. S. Joshi, "Multi-objective optimization for silicon wafer slicing using wire-EDM process," *Materials Science in Semiconductor Processing,* 39 793–806 (2015).
[4] S. K. Garg, A. Manna, and A. Jain, "Experimental investigation of spark gap and material removal rate of Al/ZrO2 (P)-MMC machined with wire EDM," *Journal of the Brazilian Society of Mechanical Sciences and Engineering,* 38(2) 481–491 (2016).
[5] D. Devarasiddappa, and M. Chandrasekaran, "PCA-GRA integrated multi response optimization of wire-cut EDM of Ti–6Al–4V alloy for sustainable production," In *Advances in Mechanical Engineering* 257–269 (2020).
[6] V. Praveen, D. R. Raju, and J. Raju, "Assessment of optimal parameters of wire EDM on Ni-coated Al2O3p/AA7075 MMCs using PCA coupled GRA," *Arabian Journal for Science and Engineering*, 46(8) 7953–7966 (2021).
[7] A. K. Rouniyar, and P. Shandilya, "Soft computing techniques for modelling and multi-objective optimization of magnetic field assisted powder mixed EDM process," *Neural Computing and Applications*, 34 1–22 (2022).
[8] S. C. Mondal, and P. Mandal, "Application of artificial neural network for modeling surface roughness in centerless grinding operation," *In Proc. 5th International and 26th all Indian Manufacturing Technology, Design and Research Conference*, 601–608 (2014).
[9] P. Mandal, and S. C. Mondal, "An application of artificial neural network and particle swarm optimisation technique for modelling and optimisation of centreless grinding process," *International Journal of Productivity and Quality Management* 20(3) 344–362 (2017).
[10] K. Wang, H. L. Gelgele, Y. Wang, Q. Yuan, and M. Fang, "A hybrid intelligent method for modelling the *EDM process," International Journal of Machine Tools and Manufacture*, 43(10) 995–999 (2003).
[11] S. Assarzadeh, and M. Ghoreishi, "Neural-network-based modeling and optimization of the electro-discharge machining process," *International Journal of Advanced Manufacturing Technology*, 39 488–500 (2008).
[12] G. Rangajanardhaa, "Development of hybrid model and optimization of surface roughness in electric discharge machining using artificial neural networks and genetic algorithm," *Journal of Materials Processing Technology*, 209(3) 1512–1520 (2009).
[13] K. P. Somashekhar, N. Ramachandran, and J. Mathew, "Optimization of material removal rate in micro-EDM using artificial neural network and genetic algorithms," *Materials and Manufacturing Processes*, 25(6) 467–475 (2010).
[14] A. Jain, S. K. Cheruku and Y. Shrivastava, "Fabrication and machining of metal matrix composite using electric discharge machining: A short review," *EVERGREEN Joint Journal of Novel Carbon Resource Sciences & Green Asia Strategy*, 08(4) 740–749 (2021).
[15] S. Sharma, K. Mausam, and K. Sharma, "Effects of nanoparticles on the MRR and TWR of graphenebased composite by electro discharge machining," *EVERGREEN Joint Journal of Novel Carbon Resource Sciences & Green Asia Strategy*, 09(04) 1021–1030 (2022).
[16] A. Srivastava, and S. K. S. Yadav, "Multi-objective optimization of milling ECSM process through PCA-based GRA," *EVERGREEN Joint Journal of Novel Carbon Resource Sciences & Green Asia Strategy*, 10 (4) 2230–2236 (2023).
[17] A. Saha, S. Chatterjee, and G. K. Bose, "Investigation of effect of CNC milling parameters on cylindricity and perpendicularity of milled circular pockets using CMM," *EVERGREEN Joint Journal of Novel Carbon Resource Sciences & Green Asia Strategy*, 09(03) 745–751 (2022).
[18] P. Saha, A. Singha, and S. K. Pal, "Soft computing models based prediction of cutting speed and surface roughness in wire electro-discharge machining of tungsten carbide

cobalt composite," *The International Journal of Advanced Manufacturing Technology*, 39(1–2) 74–84 (2008).

[19] O. I. Abiodun, A., Jantan, A. E. Omolara, K. V. Dada, N. A. Mohamed, and H. Arshad, "State-of-the-art in artificial neural network applications: A survey," *Heliyon*, 4(11), 00938 (2018).

[20] N. Nisrina, M. I. Kemal, I. A. Akbar, and T. Widianti, "The effect of genetic algorithm parameters tuning for route optimization in travelling salesman problem through general full factorial design analysis," *EVERGREEN Joint Journal of Novel Carbon Resource Sciences & Green Asia Strategy*, 09(01) 163–203 (2022).

[21] S. A. Shaedi, N. Mohd-Ghazali, J. T. Oh, R. Ahmad, and Y. Mohd-Yunos, "Entropy generation minimization of two-phase flow in a mini channel with genetic algorithm," *EVERGREEN Joint Journal of Novel Carbon Resource Sciences & Green Asia Strategy*, 6(1) 39–43 (2019). doi:10.5109/2321004.

[22] A. K. Srivastava, M. Maurya, A. Saxena, N. K. Maurya, and S. P. Dwivedi, "Statistical optimization by response surface methodology of process parameters during the cnc turning operation of hybrid metal matrix composite," *EVERGREEN Joint Journal of Novel Carbon Resource Sciences & Green Asia Strategy*, 8(1) 51–62 (2021). doi:10.5109/4372260.

[23] S. T. Hashemi, O. M. Ebadati E, and H. Kaur, "A hybrid conceptual cost estimating model using ANN and GA for power plant projects," *Neural Computing and Applications*, 31 2143–2154 (2019).

[24] G. Ghosh, P. Mandal, and S. C. Mondal, "Modeling and optimization of surface roughness in keyway milling using ANN, genetic algorithm, and particle swarm optimization," *International Journal of Advanced Manufacturing Technology* 100 1223–1242 (2019). https://doi.org/10.1007/s00170-017-1417-4

[25] M. K. Gupta, P. K. Sood, and V. S. Sharma, "Machining parameters optimization of titanium alloy using response surface methodology and particle swarm optimization under minimum quantity lubrication environment," *Manufacturing Processes*, 31(13) 1671–1682 (2016).

[26] A. K. Srivastava, M. Maurya, A. Saxena, N. K. Maurya and S. P. Dwivedi, "Statistical optimization by response surface methodology of process parameters during the CNC turning operation of hybrid metal matrix composite," *EVERGREEN Joint Journal of Novel Carbon Resource Sciences & Green Asia Strategy*, 8(1) 51–62 (2021). doi:10.5109/4372260

[27] K. Debnath, I. Singh, and A. Dvivedi, "Rotary mode ultrasonic drilling of glass fiber-reinforced epoxy laminates," *Journal of Composite Materials*, 49(8) 949–963 (2015). doi:10.1177/0021998314527857

[28] F. D. Ning, W. L. Cong, Z. J. Pei, and C. Treadwell, "Rotary ultrasonic machining of CFRP: A comparison with grinding," *Ultrasonics*, 66 125–132 (2016).

[29] R. P. Singh, and S. Singhal, "Experimental investigation of machining characteristics in rotary ultrasonic machining of quartz ceramic," *Proceedings of the Institution of Mechanical Engineers, Part L: Journal of Materials: Design and Applications*, 232(10) 870–889 (2018).

[30] V. Kumar and H. Singh, "Machining optimization in Rotary Ultrasonic Drilling of BK-7 through response surface methodology using desirability approach," *Journal of the Brazilian Society of Mechanical Sciences and Engineering*, 40(83) (1–14) (2018).

[31] D. Sindhu, L. Thakur, and P. Chandna, "Multiobjective optimization of rotary ultrasonic machining parameters for quartz glass using Taguchi-Grey Relational Analysis (GRA)," *Silicon*, 11(4) 2033–2044 (2019). doi:10.1007/s12633-018-0019-6.

[32] V. Kumar, and H. Singh, "Regression analysis of surface roughness and microstructural study in rotary ultrasonic drilling of BK7," *Ceramics International*, 44(14) 16819–16827 (2018). doi:10.1016/j.ceramint.2018.06.117.

[33] V. Kumar, and H. Singh, "Optimization of rotary ultrasonic drilling of optical glass using Taguchi method and utility approach," *Engineering Science and Technology, an International Journal*, 22(3) 956–965 (2019). doi: 10.1016/j.jestch.2019.02.004.
[34] D. Sindhu, L. Thakur, and P. Chandna, "Parameter optimization of rotary ultrasonic machining on quartz glass using response surface methodology (RSM)," *Silicon*, 12(2) 629–643 (2020). doi:10.1007/s12633-019-00160-2.
[35] B. M. A. Abdo, A. El-Tamimi, and H. Alkhalefah, "Parametric analysis and optimization of rotary ultrasonic machining of zirconia (ZrO2) ceramics," *IOP Conference Series: Materials Science and Engineering*, 727(1) 1–6 (2009).
[36] M. Maurya, N. K. Maurya and V. Bajpai, "Effect of SiC reinforced particle parameters in the development of aluminium based metal matrix composite," *EVERGREEN Joint Journal of Novel Carbon Resource Sciences & Green Asia Strategy*, 6(3) 200–206 (2019). doi:10.5109/2349295
[37] H. Sosiati, Y. A. Shofie and A. W. Nugroho, "Tensile properties of Kenaf/E-glass reinforced hybrid polypropylene (PP) composite with different fiber loading," *EVERGREEN Joint Journal of Novel Carbon Resource Sciences & Green Asia Strategy*, 5(2) 1–5 (2018).

7 Optimizing EDM Performance with Cu-MWCNT Composite Tools

A Taguchi-Based GRA Approach

Prosun Mandal and Akash Nag

7.1 INTRODUCTION

Among various non-traditional machining processes, electrical discharge machining (EDM) stands out as one of the most significant and widely utilized techniques. This process removes material by employing a series of discontinuous electric discharges, which erode the workpiece to create a replica of the tool's shape. The entire operation takes place within a dielectric fluid environment, ensuring precision and efficiency [1]. The machining performance is evaluated based on two important output responses: the rate of material removal normally termed as MRR and the rate of wear of electrode material termed as EWR. The enhancement of MRR and reduction of EWR are the basic requirements for achieving more effective, efficient, and economical EDM performance [2]. Four important controlling factors, namely, pulse-on-time, discharge current, voltage, and duty cycle, govern the performance of EDM [3]. Machining performing with the optimal parameters is one of the best ways to get better machining performance [4–8]. In EDM, a proper combination of machining parameters selection has an important role in getting maximum MRR and minimum EWR. Solving this type of multi-objective optimization problem is still a challenging job [9]. Tang and Du [10] used the Taguchi-based grey relational analysis (GRA) technique to resolve the multi-objective optimization problem for the EDM operation of titanium-based composite material. Mia et al. [11] solved the problem with the multiple objectives of turning process using the Taguchi-based GRA technique. The issue related to the single-objective optimization problem was easily overcome by this method. Das et al. [12] used the Taguchi-based GRA technique to optimize MRR and SR simultaneously. Kuram and Ozcelik [13] also successfully used Taguchi based GRA technique to obtain the optimal parameter setting for

 DOI: 10.1201/9781003769675-7

micro-milling operations. Singh [14] applied design of experiments (DOE) and the Taguchi-based GRA technique for finding optimized EDM process parameters for machining of 6061Al/Al_2O_3p/20P aluminium metal matrix composites. Huang et al. [15] used the grey Taguchi technique for the optimization of multiple performance parameters of EDM operation during machining of titanium alloy.

The machining efficiency depends on material, manufacturing method, and design of the electrode to a large extent [16]. Easy formability, high availability, and low cost are also necessary criteria for the fabrication of new tool materials [17]. The development of new tool materials is an important area of research for the EDM industry. The tool material should possess some basic properties like high thermal conductivity, high melting point, electrical conductivity, and tool wear resistance. It must have enough strength to withstand high compressive stress during machining [18]. Composite tool materials offer a unique advantage by combining the beneficial properties of two or more materials, making them an excellent choice for advanced machining applications. Among fabrication methods, electrodeposition stands out as an effective and reliable technique [19,20]. Copper (Cu), known for its exceptional electrical and thermal conductivity, is a popular choice for EDM tools. However, its low melting point leads to increased electrode erosion during machining [21]. Integrating a secondary material with superior electrical and thermal conductivity, as well as a higher melting point, can significantly enhance the tool's overall performance and durability.

Carbon nanotubes (CNTs) [22] exhibit exceptional electrical, thermal, and mechanical properties, maintaining their structural integrity even at temperatures up to 4,000 K [23]. Studies have reported significant improvements in electrical and thermal conductivity for Cu-based CNT composite coatings, fabricated using the electrodeposition method [24,25]. These advancements make CNT-enhanced composites highly suitable for advanced machining applications, including EDM. Cu-based CNT composite-coated electrodes have proven to be highly effective in EDM operations, significantly enhancing the properties of the machined surface [26–28]. Suzuki et al. [29] demonstrated the use of Cu-CNT composite electrodes fabricated via the electrodeposition method in EDM. Their study revealed a notable reduction in electrode wear ratio for Cu-CNT composite electrodes compared to conventional copper electrodes. Furthermore, rapid metallization of EDM electrodes was successfully achieved through electroplating, further enhancing their performance and longevity.

The optimal machining condition and suitable tool material play an important role in improving EDM performance. In this paper, the EDM operation was performed with a Cu-multi-walled carbon nanotube (MWCNT) composite electrode. The Taguchi L9 technique was used to prepare the design matrix for experimentation. The EDM process was optimized in two ways: first, each of the performance characteristics was optimized using Taguchi-based signal-to-noise (*S/N*) ratio; second, all performance characteristics were simultaneously optimized by applying the GRA-based Taguchi method.

7.2 EXPERIMENTATION

The electro-co-deposition method is an effective technique for fabricating metal matrix CNT composites. In this study, a copper matrix MWCNT composite-coated electrode was developed using this method. The basic electrolytic solution comprised copper (II) sulphate (200 g/L) and sulphuric acid (20 g/L), with MWCNTs (diameter: 20 nm; length: 20 μm) added at a concentration of 0.5 g/L. To address the agglomeration of MWCNTs, cetyltrimethylammonium bromide (CTAB) was employed. Achieving a uniform dispersion of MWCNTs in the electrolyte is critical, and this was accomplished using both an ultrasonic homogenizer and a magnetic stirrer. Aluminium alloy and pure copper were selected as the cathode and anode, respectively. Prior to electro-co-deposition, the aluminium alloy was anodized to ensure strong adhesive coating formation [30,31]. Figure 7.1 illustrates the microstructural image of the copper matrix MWCNT composite-coated tool surface. Compositional analysis (Figure 7.2) revealed that the surface contained 77.73% pure copper and 5.74% MWCNTs. The inclusion of MWCNTs in the copper matrix not only strengthened the composite but also enhanced its electrical and thermal conductivity [30].

The experiments were conducted on a mild steel workpiece using an EDM machine (Rantanparkhi, Model 400 × 250). The machining parameters, along with their three distinct levels, are detailed in Table 7.1. Before application in the EDM process, the fabricated composite electrode was thoroughly cleaned and polished to

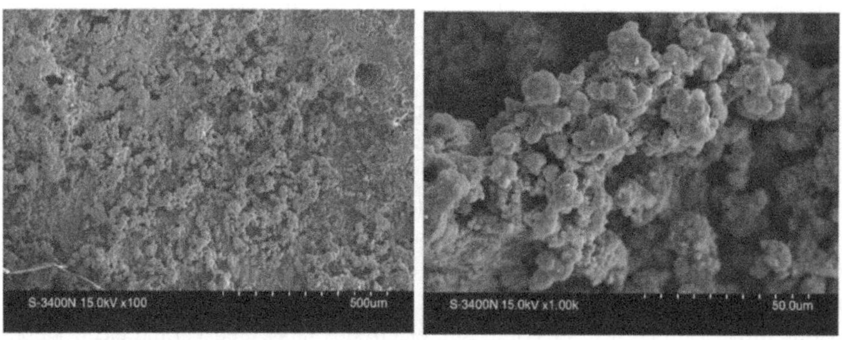

FIGURE 7.1 Surface morphology of the Cu-based MWCNT composite electrode.

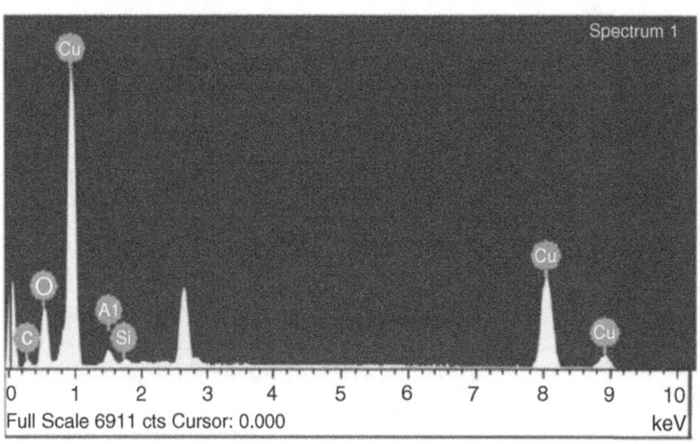

Element	Weight%
C K	5.74
O K	13.25
A1 K	2.48
Si K	0.81
Cu L	77.73
Totals	100.00

FIGURE 7.2 EDS of Cu-based MWCNT composite electrode.

TABLE 7.1
The levels of EDM process parameters

Factors	Level		
	−1	0	+1
T_{on} (μs)	12	25	38
T_{au} (%)	2	5	8
Discharge current (*I*)	3	5	7
Gap voltage (*V*)	30	40	50

TABLE 7.2
Design of experimental with evaluated output responses

Run	T_{on} (ms)	T_{au}	*I* (A)	*V*	EWR (gm/min)	MRR (mm³/min)
1	−1	−1	−1	−1	0.008333	130.9
2	−1	0	0	0	0.00551	144.26
3	−1	+1	+1	+1	0.004	157.08
4	0	−1	0	+1	0.01125	176.72
5	0	0	+1	−1	0.009474	186.02
6	0	+1	−1	0	0.005714	168.3
7	+1	−1	+1	0	0.025882	277.2
8	+1	0	−1	+1	0.015882	207.9
9	+1	+1	0	−1	0.001034	243.74

ensure a smooth surface for optimal performance. The sandpaper (800/P1500) was used for this purpose. A large number of experiments need to be carried out with three different levels of four different factors, and it is not worth practising in experimentation. The number of experiments needs to be reduced, and Taguchi L9 is one of the useful techniques in this regard. This technique was used in the present study for preparing the experimental design matrix, and experimentation was performed accordingly. Table 7.2 presents the experimental design matrix along with the corresponding evaluated output responses.

7.3 SIGNAL-TO-NOISE-BASED TAGUCHI OPTIMIZATION

Dr. Genichi Taguchi, a renowned quality scientist, has developed a robust statistical tool for designing a highly standard manufacturing system, known as the Taguchi method [32]. This method provides optimal process parameter settings with much lower variance using an orthogonal array of experiments [33]. The quality loss values, representing the deviation of a product's performance from its target functionality, are determined using the experimental data. These quality loss functions are characterized in three different ways, viz., lower-the-better, higher-the-better, and nominal-the-better [34]. Further, these loss functions are converted into a signal-to-noise (*S/N*) ratio [35]. The desired values are known as "signal" and undesired values are known as "noise". Thus, a larger *S/N* ratio expresses the better quality characteristics. This technique can be successfully utilized for optimization problem involving signal quality characteristics. It is complicated to find an optimal solution for the multi-quality features. To solve this issue, GRA is combined with the Taguchi method [36,37]. The performance characteristics used to analyse the *S/N* ratio are computed using Equations 7.1–7.3.

Lower is better (minimize):

$$\frac{S}{N} = -10\log\left(\frac{1}{n}\sum_{i=1}^{n} y_i^2\right) \tag{7.1}$$

Higher is better (maximize):

$$\frac{S}{N} = -10\log\left(\frac{1}{n}\sum_{i=1}^{n}\frac{1}{y_i^2}\right) \tag{7.2}$$

Nominal is the best:

$$\frac{S}{N} = 10\log\left(\frac{\overline{y}}{s_y^2}\right) \tag{7.3}$$

The averaging values of the observed data is denoted by y, and the variance of y is expressed as $\overline{y}$ and S_y^2 respectively. In the above mathematical expression, the number of observations is denoted by *n*. The minimization of the rate of electrode wear and the maximization of the rate of material removal from the workpiece are always desired for EDM operation. The *S/N* ratio is calculated for EWR using the "smaller-is-better" criterion (Equation 7.1) and for MRR using the "larger-is-better" criterion (Equation 7.2). Table 7.3 shows the experimental responses and corresponding *S/N* ratios.

TABLE 7.3
Calculated *S/N* ratios for EWR and MRR

Experiment no	Output responses		Signal-to-noise (*S/N*) ratios	
	EWR	MRR	EWR	MRR
1	0.008333	130.90	41.5839	42.3387
2	0.00551	144.26	45.1769	43.1829
3	0.004	157.08	47.9588	43.9224
4	0.01125	176.72	38.9769	44.9457
5	0.009474	186.02	40.4693	45.3911
6	0.005714	168.30	44.8612	44.5216
7	0.025882	277.20	31.7400	48.8558
8	0.015882	207.90	35.9819	46.3570
9	0.001034	243.74	59.7095	47.7385

The computed *S/N* ratios of EWR based on Taguchi analysis for each level of input factors are presented in Table 7.4. The influence of EDM operating parameters is identified according to the delta value. This delta value is evaluated by subtracting the minimum value of *S/N* ratio from the maximum value of *S/N* ratio for particular input factors. The highest delta value of the input factor indicates the strongest influence on the output response. The duty cycle (T_{au}) has the highest impact on EWR. Figure 7.3 is the graphical representation of Table 7.4. According to the Taguchi method, the highest value of the *S/N* ratio indicates the optimal machining condition. Figure 7.3 suggests that the low level of pulse-on-time (12 μs), high level of duty cycle (8%), medium level of discharge current (5 A), and low level of gap voltage (30 V) are the optimal machining conditions for the minimization of EWR. In a similar way, the *S/N* ratio response table (Table 7.5) for MRR reveals that the pulse-on-time denoted by T_{on} has the highest influence on MRR. The optimal machining parameters for the maximization of MRR were obtained from Figure 7.4.

TABLE 7.4
***S/N* ratio response table for EWR**

Level	T_{on}	T_{au}	I	V
1	44.91	37.43	40.81	47.25
2	41.44	40.54	47.95	40.59
3	42.48	50.84	40.06	40.97
Delta	3.47	13.41	7.9	6.66
Rank	4	1	2	3

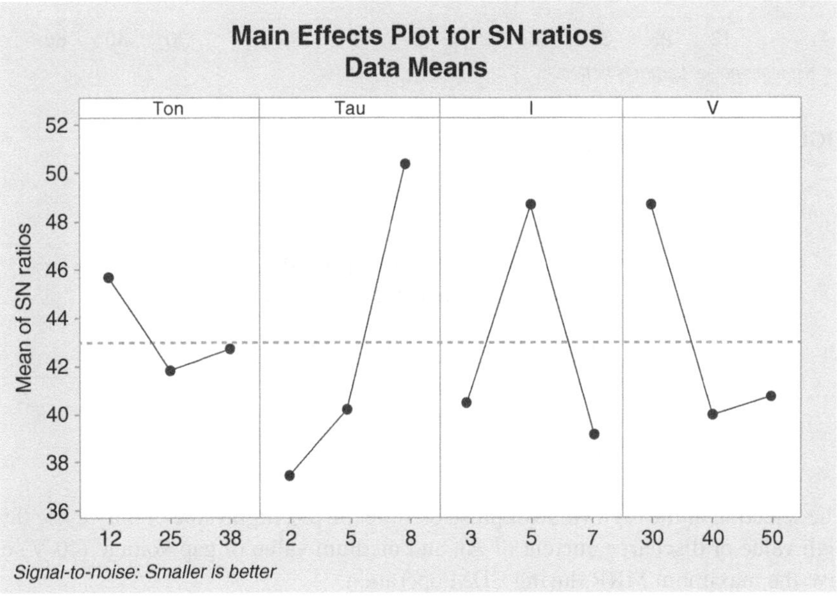

FIGURE 7.3 Main effect plot of *S/N* ratio for EWR.

TABLE 7.5
***S/N* ratio response table for MRR**

Level	T_{on}	T_{au}	I	V
1	43.15	45.38	44.41	45.16
2	44.95	44.98	45.29	45.52
3	47.65	45.39	46.06	45.08
Delta	4.5	0.42	1.65	0.45
Rank	1	4	2	3

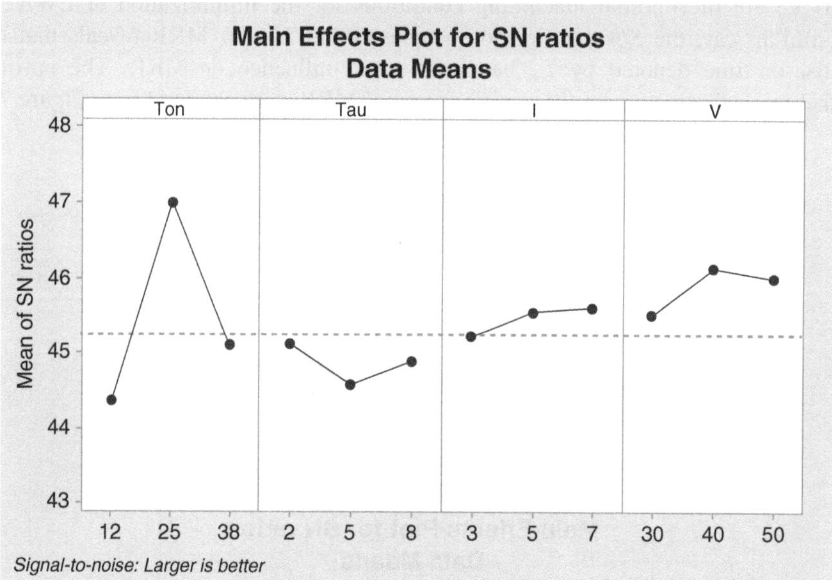

FIGURE 7.4 Main effect plot of *S/N* ratio for MRR.

TABLE 7.6
Taguchi-based EDM optimal parameters for EWR and MRR

Responses	Optimum factors			
	T_{on}	T_{au}	I	V
EWR	38 μs	8%	5 A	30 V
MRR	38 μs	8%	7 A	40 V

The selection of the high value of pulse on time (38 μs), high value of duty cycle (8%), high value of discharge current (7 A), and medium value of gap voltage (40 V) can give the maximum MRR during EDM operation.

Table 7.6 presents the responses and corresponding optimal input factors. It is clearly observed that for the enhancement of MRR and the reduction of EWR, different optimal parameter settings were obtained using the Taguchi-based single optimization technique. This issue needs to be solved by finding a single optimal setting for optimizing both MRR and EWR. Therefore, an appropriate multi-objective optimization technique is required.

7.4 GREY RRELATIONAL AANALYSIS-BASED TAGUCHI OPTIMIZATION

Multiple responses are simultaneously optimized by selecting independent input factors using GRA, and it can also overcome the discrepancy of single-objective optimization [37,38]. The complex and non-linear multi-objective problems are categorized into three different categories: first, where only maximization of the output responses is desired; second, where only minimization of the output responses is desired; and third, where both maximization of some responses and minimization of some responses are desired. In the present study, the EWR needs to be reduced, and the MRR needs to be increased. The grey relational grade is a very useful technique which can transform a multiple-objective optimization problem into a single-objective optimization problem (Kuram and Ozcelik, 2013). Four major steps are followed to accomplish the GRA.

Step 1: Normalizing of data

The GRA starts with normalizing different parameters to convert multi-scales into a unified scale. The EWR of EDM operation needs to be minimized and experimental data normalized with "the-smaller-the-better" characteristics, which is presented as follows:

$$x_i^*(k) = \frac{\max\left(x_i^k(k)\right) - x_i^0(k)}{\max\left(x_i^0(k)\right) - \min\left(x_i^0(k)\right)} \tag{7.4}$$

The MRR needs to increase and be normalized by "the-larger-the-better" characteristic and calculated as follows:

$$x_i^*(k) = \frac{x_i^k(k) - \min\left(x_i^0(k)\right)}{\max\left(x_i^0(k)\right) - \min\left(x_i^0(k)\right)} \tag{7.5}$$

where $x_i^*(k)$ is the normalized value, max $(x_i^0(k))$ is the maximum value of original experimental data, min $(x_i^0(k))$ is the least value of original experimental data of kth response, and k is 1 for EWR and 2 for MRR. Table 7.7 shows the normalized values of EWR and MRR, computed by Equations 7.4 and 7.5, respectively.

Step 2: Estimation of grey relational coefficient

The calculation of grey relational coefficients is the second step, which establishes the relation of the ideal to the experimental results. The mathematical expression for the calculation of grey relational coefficient is as follows (Equation 7.6):

$$\xi_i(k) = \frac{\Delta_{\min} - \zeta\Delta_{\max}}{\Delta_{0i} - \zeta\Delta_{\max}} \tag{7.6}$$

TABLE 7.7
Normalization and deviation of responses

	Experimental results		Normalization		Deviation	
Ex no	**EWR**	**MRR**	**EWR**	**MRR**	**EWR**	**MRR**
1	0.0083	130.90	0.7063	0	0.2937	1
2	0.0055	144.26	0.8199	0.0913	0.1801	0.9087
3	0.0040	157.08	0.8806	0.1789	0.1194	0.8211
4	0.0113	176.72	0.5889	0.3132	0.4111	0.6868
5	0.0095	186.02	0.6603	0.3768	0.3397	0.6232
6	0.0057	168.30	0.8117	0.2556	0.1884	0.7444
7	0.0259	277.20	0	1	1	0
8	0.0159	207.90	0.4026	0.5263	0.5976	0.4737
9	0.0011	243.74	1	0.7713	0	0.2287

$$0 < \xi_i(k) \leq 1 \tag{7.7}$$

where $\Delta_{oi}(k) = \left| x_0^*(k) - x_i^*(k) \right|$ and $\Delta_{max} = 1.00$, $\Delta_{min} = 0.00$

The value of the coefficient ζ lies between 0 and 1. In the current analysis, this value is chosen as 0.5 for the calculation of the grey relational coefficient. Table 7.7 shows the computed grey relational coefficients for EWR and MRR.

Step 3: Evaluation of grey relational grade

The third step of the GRA is to calculate the grade of the grey relational, which is the average value of the grey relational coefficients of responses. Its mathematical expression is as follows:

$$\alpha_i = \frac{1}{n} \sum_{k=1}^{n} \xi_i(k) \tag{7.8}$$

Where n is the number of EDM responses (in the present study, two responses are considered), the grade of the grey relational is considered as the combined response of the multiple responses that are EWR and MRR. The highest value of grey relational grade indicates the best result.

Step 4: The ranking of grey relational grade

The extreme right side column of Table 7.8 shows the grey relational ranking. The highest value of grey relational grade is assigned as rank 1, which is the best performing response, and the corresponding parameter setting is the optimal parameter for EDM operation. According to Table 7.8, experiment no 6 is assigned as rank 1, which means the selection of the medium level of pulse on time, high level of duty cycle, low level of discharge current, and a medium level of gap voltage for EDM operation leads to the minimization of EWR and maximization of MRR.

The effects of each of the process parameters on the responses are analysed by the Taguchi method. The higher-the-better (Equation 7.2) *S/N* ratio quality characteristic was adopted for the analysis of grey relational grade. Figure 7.5 presents the impact of EDM process parameters on the output responses. The A2B3C1D2

TABLE 7.8
Grey relational coefficients and grades

	Grey relational co-efficient		Grey relational grades	
Ex no	**EWR**	**MRR**	**Average value**	**Rank**
1	0.6299	0.3333	0.4924	8
2	0.7351	0.3549	0.5568	4
3	0.8073	0.3785	0.6143	2
4	0.5488	0.4213	0.4969	7
5	0.5955	0.4451	0.4986	6
6	0.7264	0.4018	0.8631	1
7	0.3333	1	0.4234	9
8	0.4556	0.5135	0.5708	3
9	1	0.6862	0.5000	5

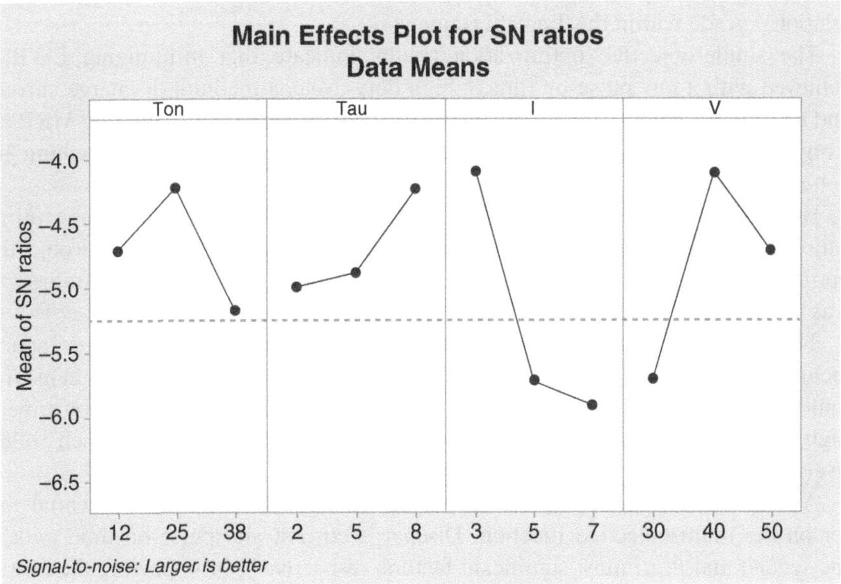

FIGURE 7.5 Process parameter effects on grey relational grade.

TABLE 7.9
S/N ratio response table for grey relational grade (GRG)

Level	T_{on}	T_{au}	I	V
1	−5.157	−6.564	−4.1	−6.073
2	−4.465	−5.333	−5.727	−4.609
3	−6.118	−3.844	−5.914	−5.059
Delta	1.653	2.72	1.814	1.463
Rank	3	1	2	4

process parameters are found to be optimal for multiple performance characteristics. Table 7.9 presents the *S/N* ratio response table for the grey relational grade. From the table, it is observed that duty cycle, which is denoted by T_{au}, has the highest influence on the combined output responses.

7.5 CONCLUSION

The selection of an appropriate tool material and optimal parameter settings plays a critical role in enhancing the performance of EDM applications. In this study, a Cu-based MWCNT composite-coated aluminium alloy tool was employed in the EDM process. Taguchi-based GRA was utilized to determine the optimal machining parameter settings.

The optimization of the EDM process was carried out in two stages. First, single-objective optimization was performed using the *S/N* ratio-based Taguchi method to optimize two key performance characteristics: EWR and MRR. Second, a multi-objective optimization of the EDM process was conducted using the grey relational grade within the Taguchi framework.

The single-objective optimization results indicate that minimizing EWR is achieved with a low pulse-on time, a high duty cycle, a medium discharge current, and a low gap voltage. Conversely, the optimal parameters for maximizing MRR are a high pulse-on time, a high duty cycle, a high discharge current, and a medium gap voltage.

However, a conflict arises between the parameter settings required for the minimization of EWR and the maximization of MRR. To resolve this issue, a multi-objective optimization approach using the grey relational grade within the Taguchi framework was employed.

The multi-objective optimization using the grey relational grade-based Taguchi technique identifies the A2B3C1D2 parameter combination as optimal for achieving multiple performance objectives. This corresponds to a medium pulse-on time, a high duty cycle, a low discharge current, and a medium gap voltage, which collectively optimize all the responses.

Among these parameters, the duty cycle emerges as the most influential factor on the multi-objective function. Discharge current and pulse-on time rank as the second and third most significant factors, respectively, in impacting the overall performance.

REFERENCES

[1] Majumder A, Das PK, Majumder A, Debnath M. An approach to optimize the EDM process parameters using desirability-based multi-objective PSO. *Prod. Manuf. Res.* 2014; 2(1):228–240.

[2] Majumder A. Process parameter optimization during EDM of AISI 316 LN stainless steel by using fuzzy based multi-objective PSO. *J. Mech. Sci. Technol.* 2013; 27(7):2143–2151.

[3] Dewangan S, Gangopadhyay S, Biswas CK. Multi-response optimization of surface integrity characteristics of EDM process using grey-fuzzy logic-based hybrid approach. *Int. J. Eng. Sci. Technol.* 2015; 18(3):361–368.

[4] Mondal SC, Mandal P. Application of artificial neural network for modeling surface roughness in centerless grinding operation. *5th Int. & 26th All India Manufacturing Technology, Design and Research Conference*, 2014, December 12–14, IIT Guwahati, Assam, India.
[5] Mondal SC, Mandal P. An application of particle swarm optimization technique for optimization of surface roughness in centerless grinding operation. In: Chakrabarti A., editor. *ICoRD'15–Research into Design Across Boundaries Volume 2. Smart Innovation, Systems and Technologies*, New Delhi: Springer; 2003. p. 687–697.
[6] Ghosh G, Mandal P, Mondal SC. Modeling and optimization of surface roughness in keyway milling using ANN, genetic algorithm, and particle swarm optimization. *Int. J. Adv. Manuf. Technol.* 2019; 100(5):1223–1242.
[7] Mandal P, Mondal SC. An application of artificial neural network and particle swarm optimisation technique for modelling and optimisation of centreless grinding process. *Int. J. Product. Qual. Manag.* 2017; 20(3):344–362.
[8] Mandal P, Mondal SC. Multi-objective optimization of Cu-MWCNT composite electrode in electro discharge machining using MOPSO-TOPSIS. *Measurement.* 2021; 169:108347.
[9] Mandal D, Pal SK, Saha P. Modeling of electrical discharge machining process using back propagation neural network and multi-objective optimization using non-dominating sorting genetic algorithm-II. *J. Mater. Process. Technol.* 2007; 186(1–3):154–162.
[10] Tang L, Du YT. Multi-objective optimization of green electrical discharge machining Ti–6Al–4V in tap water via Grey-Taguchi method. *Mater. Manuf. Process.* 2014; 29 (5):507–513.
[11] Mia M, Khan MA, Rahman SS, Dhar NR. Mono-objective and multi-objective optimization of performance parameters in high pressure coolant assisted turning of Ti-6Al-4V. *Int. J. Adv. Manuf. Technol.* 2017; 90(1–4):109–118.
[12] Das MK, Kumar K, Barman TK, Sahoo P. Optimisation of EDM process parameters using grey-Taguchi technique. *Int. J. Mach. Mach. Mater.* 2014; 15(3–4):235–262.
[13] Kuram E, Ozcelik B. Multi-objective optimization using Taguchi based grey relational analysis for micro-milling of Al 7075 material with ball nose end mill. *Measurement.* 2013; 46(6):1849–1864.
[14] Singh S. Optimization of machining characteristics in electric discharge machining of 6061Al/Al 2 O 3 p/20P composites by grey relational analysis. *Int. J. Adv. Manuf. Technol.* 2012; 63(9):1191–1202.
[15] Huang CH, Yang AB, Hsu CY. The optimization of micro EDM milling of Ti–6Al–4V using a grey Taguchi method and its improvement by electrode coating. *Int. J. Adv. Manuf. Technol.* 2018; 96(9–12):3851–3859.
[16] Jha B, Ram K, Rao M. An overview of technology and research in electrode design and manufacturing in sinking electrical discharge machining. *J. Eng. Sci. Technol. Rev.* 2011; 4(2):118–130.
[17] Beri N, Maheshwari S, Sharma C, Kumar A. Technological advancement in electrical discharge machining with powder metallurgy processed electrodes: A review. *Mater. Manuf. Process.* 2010; 25(10):1186–1197.
[18] Zaw HM, Fuh JY, Nee AY, Lu L. Formation of a new EDM electrode material using sintering techniques. *J. Mater. Process. Technol.* 1999; 89:182–186.
[19] Li L, Feng L, Bai X, Li ZY. Surface characteristics of Ti–6Al–4V alloy by EDM with Cu–SiC composite electrode. *Appl. Surf. Sci.* 2016; 388:546–550.
[20] Mandal P, Mondal SC. Investigation on the performance of Copper-Coated 6061 aluminium alloy electrode in electric discharge machining. In: Chakrabarti A., editor. *Research into Design for a Connected World.* Singapore: Springer; 2019. p. 345–355.
[21] Li L, Niu ZW, Zheng GM. Ultrasonic electrodeposition of Cu–SiC electrodes for EDM. *Mater. Manuf. Process.* 2016; 31(1):37–41.

[22] Iijima S. Helical microtubules of graphitic carbon. *Nature*. 1991; 354(6348):56–58.
[23] Miyamoto Y, Berber S, Yoon M, Rubio A, Tománek D. Onset of nanotube decay under extreme thermal and electronic excitations. *Phys. B: Condens. Matter.* 2002; 323(1–4):78–85.
[24] Mandal P, Mondal SC. Enhancement of electro-thermal and mechanical properties for Cu-SWCNT coated 6061Al. *Surf. Eng.* 2020; 36(2):135–143.
[25] Mandal P, Mondal SC. Investigation of electro-thermal property of Cu-MWCNT-coated 316L stainless steel. *Surf. Eng.* 2018; 34(9):697–704.
[26] Mandal P, Mondal SC. Development and application of Cu-SWCNT nanocomposite–coated 6061Al electrode for EDM. *Int. J. Adv. Manuf. Technol.* 2019; 103(5):3067–3076.
[27] Mandal P, Mondal SC. Surface characteristics of mild steel using EDM with Cu-MWCNT composite electrode. *Mater. Manuf. Process.* 2019; 34(12):1326–1332.
[28] Mandal P, Mondal SC. Performance analysis of copper-based MWCNT composite coated 316L SS tool in electro discharge machining. *Mach. Sci. Technol.* 2021; 25(3):422–437.
[29] Suzuki T, Kato M, Saito H, Iizuka H. Effect of carbon nanotube (CNT) size on wear properties of Cu-based CNT composite electrodes in electrical discharge machining. *J. Solid Mech. Mater. Eng.* 2011; 5(7):348–359.
[30] Mandal P, Chatterjee S, Chakraborty S. (2024). Bi-objective optimization of an EDM process with Cu-MWCNT composite tool using single-valued neutrosophic grey relational analysis. *World J Eng.* 22(3): 472–481. https://doi.org/10.1108/WJE-10-2023-0443
[31] Mandal P, Mondal SC. Investigation of Electro-Thermal property for Cu-MWCNT composite coating on anodized 6061 aluminium alloy. *Appl. Surf. Sci.* 2018; 454:138–147.
[32] Taguchi G. *Introduction to Quality Engineering*. Tokyo, Japan: Asian Productivity Organization (APO); 1990.
[33] Ganguly D, Acherjee B, Kuar AS, Mitra S. Hole characteristics optimization in Nd: YAG laser micro-drilling of zirconium oxide by grey relation analysis. *Int. J. Adv. Manuf. Technol.* 2012; 61(9–12):1255–1262.
[34] Asiltürk I, Akkuş H. Determining the effect of cutting parameters on surface roughness in hard turning using the Taguchi method. *Measurement.* 2011; 44(9):1697–1704.
[35] Nalbant M, Gökkaya H, Sur G. Application of Taguchi method in the optimization of cutting parameters for surface roughness in turning. *Mater. Des.* 2007; 28(4):1379–1385.
[36] Rathinasuriyan C, Kumar VS. Optimisation of submerged friction stir welding parameters of aluminium alloy using RSM and GRA. *Adv. Mater. Process. Technol.* 2020;7(4):696–709.
[37] Mandal P. Data-driven optimization of manufacturing processes. *IGI Global; 2021, Chapter 4, Optimization of Surface Roughness in Centreless Grinding Process Based on Taguchi Method*; p. 37–47.
[38] Mia M, Al Bashir M, Khan MA, Dhar NR. Optimization of MQL flow rate for minimum cutting force and surface roughness in end milling of hardened steel (HRC 40). *Int. J. Adv. Manuf. Technol.* 2017; 89(1–4):675–690.

8 Smart Machining Choices
A Hybrid CIMAS-PIV Model for Titanium Precision

Mrityunjoy Roy, Smita, Amresh Kumar Sinha, and Rishi Dwivedi

8.1 INTRODUCTION

The rapid evolution of technology has fuelled remarkable progress in materials science and manufacturing, creating a pressing need for machining processes that can deliver both precision and efficiency for increasingly complex applications. Titanium, with its exceptional strength-to-weight ratio, corrosion resistance, and biocompatibility, has emerged as a material of significant interest across diverse industries, including aerospace, biomedical engineering, and military. However, these desirable properties also make titanium notoriously difficult to machine using traditional methods.

Conventional machining techniques often struggle with titanium due to its low thermal conductivity and high chemical reactivity. The low thermal conductivity leads to localized heat buildup at the cutting interface, accelerating tool wear, and reducing tool life. Additionally, titanium's tendency to react with cutting tool materials causes adhesion and material transfer, compromising precision and surface quality. These challenges not only impact efficiency but also increase costs, hindering the wider adoption of titanium in cost-sensitive applications. Consequently, there is a growing need to explore alternative machining processes specifically designed to address the unique challenges posed by titanium.

Non-traditional machining (NTM) processes offer a promising solution. These processes utilize innovative techniques, such as thermal, chemical, and electrochemical energy, to remove material, effectively bypassing the limitations of mechanical cutting. Examples include laser beam machining (LBM), plasma arc machining (PAM), and electrochemical machining (ECM), among others. NTM processes enable the precise machining of titanium and other difficult-to-process materials, often with reduced tool wear and improved surface quality. However, the wide array of available NTM techniques presents a new challenge: selecting the optimal process for a given application. This decision is further complicated by the need to balance multiple, often conflicting, criteria, such as material removal rate (MRR), geometric tolerances, surface finish, energy consumption, and cost.

DOI: 10.1201/9781003769675-8

Choosing the right NTM process for titanium machining requires a comprehensive evaluation of these diverse criteria. Current approaches often rely on multi-criteria decision-making (MCDM) techniques, which provide a structured framework for assessing and prioritizing alternative machining processes. However, many existing MCDM methods have limitations, particularly in their ability to incorporate expert judgement and address the practical relevance of decision criteria. They often prioritize quantitative data while overlooking valuable qualitative insights from experienced industry professionals. This disconnect highlights the need for a more robust methodological framework that bridges the gap between theoretical decision models and real-world manufacturing needs.

This study addresses this gap by proposing an integrated framework that combines the Criteria Importance Assessment (CIMAS) method with the Proximity Indexed Value (PIV) technique. CIMAS incorporates expert judgement into the decision-making process, allowing for a more nuanced evaluation of criteria weights. In the context of titanium machining, where experience and expertise are crucial, CIMAS provides a systematic way to assess and prioritize various machining factors. The six key criteria considered in this study are MRR, shape feature complexity, work material type, tolerance and surface finish, power requirement, and cost. Each criterion is evaluated with a focus on its practical significance, ensuring alignment with real-world manufacturing priorities.

Once the criteria weights are established using CIMAS, the PIV method is employed to rank various NTM processes, including abrasive jet machining (AJM), ultrasonic machining (USM), chemical machining (CHM), electron beam machining (EBM), LBM, ECM, electro discharge machining (EDM), and PAM. PIV enhances the decision-making process by systematically evaluating the performance of each NTM process against the weighted criteria. By integrating CIMAS and PIV, this study offers a comprehensive and adaptable framework for process selection, addressing both qualitative and quantitative aspects of decision-making.

Applying the CIMAS-PIV framework to titanium machining reveals several key findings. PAM emerges as the most suitable NTM process, followed by LBM. PAM's superior performance can be attributed to its ability to achieve high MRRs and process complex geometries with minimal thermal damage. The analysis also highlights the importance of shape feature complexity as a key criterion, reflecting the critical role of geometric precision in titanium machining. Conversely, cost is found to be less significant, underscoring the emphasis on performance and precision over economic considerations in high-value applications.

This research has significant implications for both industry and research. For manufacturers, the CIMAS-PIV framework provides a structured, evidence-based approach to process selection, facilitating informed decision-making and optimizing resource utilization. The framework's adaptability makes it a valuable tool for various applications, including the machining of other challenging materials. From a research perspective, this study contributes to the field of MCDM by introducing a hybrid framework that integrates expert judgement with advanced ranking techniques. By bridging the gap between theory and practice, this study lays the foundation for continued innovation in manufacturing decision-making tools.

8.2 LITERATURE REVIEW

A comprehensive two-phase decision model, leveraging Data Envelopment Analysis (DEA), has been introduced to identify and rank the most efficient NTM processes using multi-attribute decision-making principles (Sadhu & Chakraborty, 2011). Furthermore, a decision support framework incorporating Visual Basic 6.0 and quality function deployment was developed to streamline the selection of NTM processes (Prasad & Chakraborty, 2014). In addition, an innovative hybrid methodology integrating Factor Relationship (FARE) and Multi-Attributive Border Approximation Area Comparison (MABAC) was proposed to evaluate NTM processes effectively (Chatterjee et al., 2017). The Analytic Hierarchy Process (AHP) has also been employed to prioritize NTM processes and their associated criteria, highlighting parameters such as voltage, current, ions, and ionized particles as critical factors (Talib & Asjad, 2019). A further advancement was made by Chakraborty et al. (2020), who utilized the rough MABAC approach to address the complexities of NTM process evaluation, incorporating expert judgements to assess machining capabilities in real-world contexts. Roy et al. (2021) conducted an in-depth study on NTM process selection, combining AHP to assess the relative importance of process attributes with quality function deployment to evaluate alternatives. Their findings revealed that ECM is superior in terms of production time, corner radii, surface finish, and dimensional tolerance. Kumari and Acherjee (2022) applied the criteria importance through intercriteria correlation (CRITIC) and combinative distance-based assessment methods to address the challenging task of selecting an optimal non-conventional machining process for high-strength materials. Their analysis identified PAM as the most effective choice for machining titanium, based on performance metrics such as MRR, tolerance, surface finish, and cost-effectiveness, with their conclusions supported by prior research. Lastly, Sampath et al. (2023) demonstrated the application of the technique for order preference by similarity to ideal solution (TOPSIS) method for optimizing micromachining parameters. They resolved conflicting response variables and provided a detailed case study on EDM, including process steps, flowchart illustrations, and interpretation of results.

Khan et al. (2019) proposed the PIV method as a simpler alternative for MCDM, showcasing its reliability and rank stability in evaluating and ranking E-learning platforms compared to established techniques such as AHP, VlseKriterijuska Optimizacija I Komoromisno Resenje, and complex proportional assessment. Biswas and Anand (2020) extended the Logistics Performance Index (LPI) framework by integrating ICT adoption and CO_2 intensity, applying an advanced MCDM approach, including the PIV method, to analyse logistics competitiveness between G7 and BRICS nations, ultimately highlighting the G7's superior performance. Trung (2021) utilized a hybrid MCDM model incorporating Taguchi, Entropy, weighted aggregated sum product assessment, and PIV methods to optimize surface roughness and MRR during the external cylindrical grinding of 65G steel, underscoring PIV's utility in determining ideal grinding parameters. Goswami et al. (2022) employed a combined MEREC-PIV approach to evaluate renewable energy sources in India, assessing six key factors to identify hydroelectric power as the most suitable option under the country's climatic conditions. Saluja et al. (2023) proposed an improved

Proximity Indexed Value (IPIV) method to address the rank reversal problem in MCDM, using a MATLAB simulation-based approach to demonstrate its effectiveness in reducing rank reversal to nearly zero, with application to real-world problems. Nguyen et al. (2024) used the PIV method to select the optimal material for crankshaft manufacturing from four steel types, demonstrating that S48C steel ranked the highest regardless of the weighting method applied.

Bošković et al. (2023) proposed a novel CIMAS method for multi-criteria group decision-making in supplier selection, combining subjective expert judgement with the objective CRITIC method, and identifying on-time distribution as the most important criterion. Kara et al. (2024a) introduced a single-valued neutrosophic MCDM hybrid model for financial performance analysis of technology companies, offering a robust and generalizable decision support model that provides valuable insights for financial managers. Kara et al. (2024b) developed a Picture Fuzzy CIMAS-ARTASI model for website performance analysis in Human Resource Management, offering an advanced hybrid MCDM decision support system for effective executive recruitment, with implications for selecting the most suitable e-recruitment site for appointing a chief information security officer. Kara et al. (2024c) formulated an interval-valued spherical fuzzy criteria importance assessment (CIMAS)-integrated simple weighted sum-product hybrid model for blockchain platform selection in digital projects, offering a robust and generalizable group decision-analytic tool with novel insights and implications for project managers and practitioners. Bouraima et al. (2024) introduced a novel hybrid method, combining the CIMAS and Alternative Ranking Order Method Accounting for Two-Step Normalization (AROMAN) techniques, to assess and formulate healthcare referral system strategies in developing countries, identifying infrastructure improvement as the most effective strategy to address key challenges like inadequate facilities and health information systems.

While significant progress has been made in research on NTM processes, critical gaps persist in methodologies for selecting the most appropriate process for machining difficult-to-handle materials like titanium. Although advanced MCDM techniques have been widely utilized, the potential synergy between the PIV method and the CIMAS model remains unexplored. This integration introduces an innovative approach by combining the practical expertise captured through the CIMAS model with the PIV method's robust ranking capabilities, allowing for a comprehensive evaluation of alternatives based on multiple criteria. The proposed framework exhibits strong rank correlation with established methods, reinforcing its credibility and positioning it as a valuable advancement for addressing the complex trade-offs involved in machining high-performance materials like titanium.

8.3 METHODOLOGY

8.3.1 CIMAS Method

Bošković et al. (2023) presented the CIMAS method, an innovative approach to MCDM that emphasizes the integration of expert knowledge and experience into the decision-making process. The methodology is built on the premise that experts with greater experience in a specific domain are better positioned to evaluate the relative significance of decision criteria accurately.

Unlike conventional MCDM methods, CIMAS directly incorporates expert judgements into the weighting process by assigning weights based on individual experience levels, which are then aggregated using a weighted averaging technique. This ensures that the final criteria weights represent a well-rounded and informed perspective.

The adaptability of CIMAS is highlighted through its application in various decision-making contexts, such as choosing suitable NTM processes. In this domain, CIMAS supports manufacturers in evaluating and prioritizing critical factors like precision, surface quality, and cost, enabling a more informed and effective selection of machining techniques.

At its core, CIMAS offers a clear and structured framework for integrating expert judgement into MCDM. By harnessing the insights and expertise of experienced professionals, this approach improves the accuracy and dependability of criteria weighting, fostering more assured and impactful decision-making.

Step 1: Criteria Delineation

The first step in any effective decision-making process is to establish clear and objective evaluation criteria. This requires a careful review of existing literature and/or engaging with subject-matter experts to identify the appropriate benchmarks against which the importance of the decision will be evaluated.

Step 2: Expert Panel Constitution

The decision-maker needs to carefully determine the right number of experts to include in the evaluation process. This decision should be based on a deep understanding of the field and involve collecting information about the professional experience of potential contributors.

Step 3: Expertise Quantification

A thorough evaluation of each expert's experience is crucial. Experts are asked to offer detailed accounts of their time within the relevant domain, often measured in years of service. Based on this information, the decision-maker assigns a weight to each expert, reflecting the importance of their expertise.

Step 4: Weighting Schema

The importance of each expert's input is determined using a weighted average system. For example, if there are 'q' experts in a panel and Expert 1 has eight years of experience, Expert 2 has five years, and Expert q has seven years, their weights would be calculated as 8/20, 5/20, and 7/20, respectively. This method ensures that the opinions of those with more experience hold greater weight in the final decision. A formal mathematical formula (e.g. Equation 8.1) is used to consistently and transparently apply this weighting scheme across various decision-making scenarios.

$$W^{E_i} = \frac{E_i}{\sum_{i=1}^{q} E_i} \quad i = 1, 2, \ldots, q \tag{8.1}$$

Step 5: Formulating Input Data Matrix Based on Experts' Assessments (1–10 Scale):

At this stage, a decision-making matrix is created based on a 1–10 point scale. Each expert evaluates the importance of each criterion by assigning a score, where 10 indicates the highest level of importance and 1 the lowest. The structure of this input data matrix is outlined in Table 8.1, where E_i represents one of the q experts engaged in decision-making, C_j signifies one of the p-considered evaluation criteria, x_{ij} denotes the assessment by expert i of the importance of criterion j (on a scale of 1–10), and the values from W^{E_1} to W^{E_q} represent the weights of the experts.

Experts/Critieria	C_1	C_2	$\cdots$	C_j	$\cdots$	C_p	Expert's Weight
E_1	x_{11}	x_{12}	$\cdots$	$\cdots$	$\cdots$	x_{1p}	W^{E_1}
E_2	x_{21}	x_{22}	$\cdots$	$\cdots$	$\cdots$	x_{2p}	W^{E_2}
E_i	$\cdots$	$\cdots$	$\cdots$	x_{ij}	$\cdots$	$\cdots$	$\cdots$
E_q	x_{q1}	x_{q2}	$\cdots$	$\cdots$	$\cdots$	x_{qp}	W^{E_q}

Table 8.1 Input data matrix

Step 6: Normalize the Input Data Matrix

After constructing the input data matrix, the next step is normalization. This process scales the data values to fall within the range of 0–1, simplifying and enhancing the decision-making process. The normalization is performed using Equation 8.2, with the results illustrated in Table 8.2.

$$x_{ij}^{*} = \frac{x_{ij}}{\sum_{i=1}^{q} x_{ij}}, \; i = 1, 2, \ldots., p \tag{8.2}$$

Experts/Critieria	C_1	C_2	C_j	C_p
E_1	x_{11}^{*}	x_{12}^{*}	$\cdots$	x_{1p}^{*}
E_2	x_{21}	x_{22}	$\cdots$	x_{2p}
E_i	$\cdots$	$\cdots$	x_{ij}	$\cdots$
E_q	x_{q1}	x_{q2}	$\cdots$	x_{qp}

Table 8.2 Normalized input-data matrix

Step 7: Calculate the Expert-Weighted Matrix

In this step, each value in the normalized input data matrix is multiplied by the corresponding expert's weight determined in Step 4. This operation, performed using Equation 8.3, produces the expert-weighted matrix, as demonstrated in Table 8.3.

$$\hat{x}^{*}_{ij} = x_{ij}^{*} \cdot W^{E_i}, i = 1, 2, \ldots, q; j = 1, 2, \ldots, p \tag{8.3}$$

Experts/Critieria	C_1	C_2	$\cdots$	C_j	$\cdots$	C_p
E_1	$\widehat{x^*}_{11}$	$\widehat{x^*}_{12}$	$\cdots$	$\cdots$	$\cdots$	$\widehat{x^*}_{1p}$
E_2	$\widehat{x^*}_{21}$	$\widehat{x^*}_{22}$	$\cdots$	$\cdots$	$\cdots$	$\widehat{x^*}_{2p}$
E_i	$\cdots$	$\cdots$	$\cdots$	$\widehat{x^*}_{ij}$	$\cdots$	$\cdots$
E_q	$\widehat{x^*}_{q1}$	$\widehat{x^*}_{q2}$	$\cdots$	$\cdots$	$\cdots$	$\widehat{x^*}_{qp}$

Table 8.3 Expert-weighted matrix

Step 8: Determine the Maximum and Minimum Values of Each Criterion in the Expert-Weighted Matrix

The main objective of this stage is to ascertain the maximum ($R_{j\max}$) and minimum value ($R_{j\min}$) of each criterion by columns. This calculation is executed using Equations 8.4 and 8.5, as demonstrated in Table 8.4.

$$R_{j\max} = \max_{i} \widehat{x}^*_{ij},\ j = 1,2,\ldots.,p, \tag{8.4}$$

$$R_{j\min} = \min_{i} \widehat{x}^*_{ij},\ j = 1,2,\ldots.,p, \tag{8.5}$$

Experts/Critieria	C_1	C_2	$\cdots$	C_j	$\cdots$	C_p
E_1	$\widehat{x^*}_{11}$	$\widehat{x^*}_{12}$	$\cdots$	$\cdots$	$\cdots$	$\widehat{x^*}_{1p}$
E_2	$\widehat{x^*}_{21}$	$\widehat{x^*}_{22}$	$\cdots$	$\cdots$	$\cdots$	$\widehat{x^*}_{2p}$
E_i	$\cdots$	$\cdots$	$\cdots$	$\widehat{x^*}_{ij}$	$\cdots$	$\cdots$
E_q	$\widehat{x^*}_{q1}$	$\widehat{x^*}_{q2}$	$\cdots$	$\cdots$	$\cdots$	$\widehat{x^*}_{qp}$
$R_{j\max}$	$R_{1\max}$	$R_{2\max}$	$\equiv$	$R_{j\max}$	$\equiv$	$R_{p\max}$
$R_{j\min}$	$R_{1\min}$	$R_{2\min}$	$\equiv$	$R_{j\min}$	$\equiv$	$R_{p\min}$

Table 8.4 Maximum and minimum values of each criterion in the expert-weighted matrix

Step 9: Compute the Difference between the Minimum and Maximum Values

In this phase, the difference (B_j) is calculated by subtracting the minimum value from the maximum value derived in the previous step, using Equation 8.6.

$$B_j = R_{j\max} - R_{j\min},\ j = 1,2,..,p \tag{8.6}$$

Step 10: Use the Final Ranking Formula

In this step, the importance of criteria (L_j) is determined by employing the final ranking equation, i.e. Equation 8.7.

$$L_j = \frac{B_j}{\sum_{j=1}^{p} B_j}, j = 1,2,\ldots,p \tag{8.7}$$

Step 11: Checking the Reliability Index (RI)

This step involves conducting a second round of interviews with experts, who remain unaware of the results from the initial round regarding the assessed criteria. In this round, experts evaluate the importance of each criterion on a scale from 0% to 100%, ensuring that the total importance assigned across all 'n' criteria equals 100%. The outcomes from both rounds are then compared to assess the reliability of the results. If denoted as P_j for the responses obtained in the second round of interviews and L_j for the previously acquired CIMAS weights, the reliability index (RI) can be computed employing Equation 8.8.

$$\text{RI} = \frac{\sum_{j=1}^{n} \left| L_j * 100 - P_j \right|}{100} \tag{8.8}$$

If the RI is below 0.1, it suggests that the results are considered reliable. However, if the RI exceeds this value, it is recommended that experts repeat the criteria assessment process.

8.3.2 PIV Method

The PIV method is a versatile MCDM technique used to evaluate and rank alternatives based on multiple criteria. Its main goal is to establish a systematic and dependable decision-making process while addressing common challenges like rank reversal, which are often encountered with traditional methods such as TOPSIS and AHP. The PIV method achieves this by calculating the proximity of each alternative to an ideal solution using weighted and normalized values, resulting in logical and consistent rankings. The methodology is adaptable, suitable for both benefit and cost criteria, and features a straightforward normalization and weighting process. PIV has been successfully applied across various domains, including material selection, process optimization, and site allocation, demonstrating its effectiveness in managing complex decision-making problems with minimal susceptibility to inconsistencies.

The PIV method is applied to rank m alternatives (A_1, A_2,....,A_m) based on n criteria (C_1,C_2,...,C_n). The approach is divided into the following steps.

Step 1: Define the Decision Problem and Construct the Decision Matrix

The decision problem is defined by specifying the alternatives and the criteria. The decision matrix D is formed with elements y_{ij}, where y_{ij} represents the performance of alternative A_i on criterion C_j.

Step 2: Normalize the Decision Matrix

The normalization step ensures that all criteria values are converted into a dimensionless form. The normalized value r_{ij} for the decision matrix is calculated using the following formula:

$$r_{ij} = \frac{y_{ij}}{\sqrt{\sum_{i=1}^{m} y_{ij}^2}} \tag{8.9}$$

Where

- y_{ij} is the original value of the ith alternative for the jth criterion.
- m is the number of alternatives.

Step 3: Compute Weighted Normalized Values

The weighted normalized values v_{ij} are then calculated as:

$$v_{ij} = w_j \,.\, r_{ij} \tag{8.10}$$

Step 4: Calculate Weighted Proximity Indexes

The proximity index is computed to determine the proximity of each alternative to the ideal values.

For benefit criteria:

$$u_{ij} = v_{ij,\max} - v_{ij} \tag{8.11}$$

For cost criteria:

$$u_{ij} = v_{ij} - v_{ij,\min} \tag{8.12}$$

Step 5: Determine the Proximity Index

The proximity index d_i of each alternative A_i is calculated by summing the weighted proximity indexes:

$$d_i = \sum_{j=1}^{n} u_{ij} \tag{8.13}$$

Step 6: Rank the Alternatives

The alternatives are ranked based on their d_i values in ascending order. The alternative with the smallest d_i is considered the best.

8.4 APPLICATION OF INTEGRATED CIMAS-PIV MODEL FOR NTM SELECTION

The inherent properties of titanium, such as its high strength and low thermal conductivity, present significant challenges for conventional machining techniques. These properties can lead to issues such as poor surface integrity, rapid tool wear, and reduced productivity. NTM processes, however, offer a promising alternative, capable of delivering high precision and excellent surface quality in titanium machining.

This study aims to determine the most appropriate NTM process for machining titanium using a hybrid MCDM approach. A thorough assessment is conducted on eight prominent NTM processes, considering six key machining criteria identified by Chakladar et al. (2009). These criteria include MRR, surface finish, dimensional accuracy, tool wear, process cost, and environmental impact. The hybrid CIMAS-PIV model combines various decision-making techniques to analyse the trade-offs among these criteria, thereby identifying the optimal NTM process for enhancing efficiency and quality in titanium machining. The evaluation includes the following NTM methods:

1. AJM: AJM utilizes a high-speed stream of abrasive particles propelled by pressurized gas, such as air or nitrogen, to erode the surface of the workpiece. Material removal is accomplished through the kinetic energy generated by the impacting particles, leading to micro-cutting and brittle fracture of the material.
2. USM: USM involves a tool that vibrates at ultrasonic frequencies, working in tandem with an abrasive slurry. This combination results in micro-chipping and erosion of the workpiece material through the abrasive particles and the ultrasonic vibrations.
3. CHM: CHM is a subtractive manufacturing process that utilizes controlled chemical etching to selectively remove material from the workpiece. Etchants, such as acids or alkalis, interact with the material to facilitate the removal process, while maskants protect designated areas from the chemical reactions.
4. EBM: EBM employs a focused, high-speed electron beam within a vacuum environment. When the electron beam strikes the workpiece, the kinetic energy is rapidly converted into thermal energy, resulting in the localized melting and vaporization of the material.
5. LBM: LBM employs a focused laser beam that melts and vaporizes material in a highly precise and controlled manner. The concentrated energy and high intensity of the laser enable accurate and targeted material removal, allowing for fine, intricate cuts and precise modifications to the workpiece.
6. ECM: It involves the anodic dissolution process. A specially shaped electrode and the workpiece are submerged in an electrolyte solution, through which a controlled electric current is applied. This process leads to the precise dissolution of the workpiece material in a well-regulated manner.
7. EDM: It utilizes controlled electrical discharges between a tool electrode and the workpiece submerged in a dielectric fluid. The rapid and localized heating generated by these discharges leads to material removal through melting and vaporization.
8. PAM: PAM employs a high-temperature plasma jet to melt and expel material from the workpiece. This plasma, generated by passing a gas through an electric arc, creates a high-velocity jet capable of efficient and rapid material removal.

Each of these NTM processes exhibits unique characteristics, advantages, and limitations, making them appropriate for applications in the machining of titanium and its alloys. This study seeks to thoroughly evaluate these processes to identify the most effective approach for achieving desired machining outcomes in titanium.

Selecting the most suitable NTM process for titanium requires a thorough evaluation based on critical performance criteria. In this study, we utilize these criteria to assess and compare the effectiveness of the chosen NTM processes.

1. Material Removal Rate (MRR) measures the volume of material removed from the workpiece per unit of time. It is a vital indicator of the efficiency and productivity of the machining process. Higher MRR values typically lead to shorter machining times and increased throughput.
2. Shape Feature (*F*) assesses the ability of an NTM process to produce specific geometric features and intricate shapes on the workpiece. This criterion highlights the process's versatility and adaptability in achieving complex geometries.
3. Work Material Type (*M*) evaluates the compatibility of an NTM process with the specific workpiece material. It examines the process's effectiveness in machining titanium by considering factors such as material hardness, thermal properties, and machinability.
4. Tolerance and Surface Finish (TSF) refer to the two key aspects of machining quality—dimensional accuracy and surface integrity. This criterion assesses the ability of an NTM process to achieve precise tolerances and produce the desired surface finish on the titanium workpiece.
5. Power Requirement (PR) measures the energy consumption of an NTM process. This criterion evaluates the process's energy efficiency and its implications for operational costs and environmental sustainability.
6. Cost (*C*) refers to the financial implications of an NTM process. It includes both initial capital investment and ongoing expenses such as equipment costs, tooling, consumables, maintenance, and operational costs. This criterion assesses the overall economic feasibility of the process.

The six criteria offer a comprehensive framework for evaluating NTM processes in titanium machining. By carefully examining the performance of each process against these criteria, this study aims to identify the most suitable NTM solution for achieving the desired machining outcomes with improved efficiency and quality.

8.4.1 Determining Criteria Weights with the CIMAS Method

The first step in selecting an NTM process involves a thorough assessment of the relative importance of six predefined criteria. This critical phase forms the basis for the selection process by establishing a weighted framework for comparing various NTM processes. To accomplish this, the CIMAS method is utilized — a systematic approach tailored for objective criteria weighting.

In this context, the CIMAS method leverages expert knowledge to quantify the significance of each criterion. A panel of five experts, all possessing specialized knowledge and experience in NTM processes, was assembled for this purpose. Each expert independently evaluated the six criteria, offering their subjective judgements on their relative importance. These expert opinions are systematically recorded in an input data matrix, as shown in Table 8.1. This matrix serves as the primary data source for the CIMAS method, capturing the collective insights of the expert panel and facilitating a quantitative analysis of criteria significance.

Following the CIMAS methodology for criteria weight determination, the next step involves converting the raw expert judgements into a weighted normalized input data matrix. The initial input data matrix, which contains the raw scores provided by the experts, is first normalized. This normalization process standardizes the data, ensuring that all criteria are evaluated on a comparable scale, and helps to eliminate any bias stemming from differing measurement scales or individual expert tendencies. The normalized data matrix is then multiplied by pre-determined expert weights, which reflect the relative influence of each expert's opinion based on factors such as qualifications and experience. This weighting process ensures that the final criteria weights accurately capture the collective expertise of the panel. The resulting weighted normalized input data matrix, as illustrated in Table 8.2, provides a refined and objective representation of criteria significance.

In the CIMAS weight calculation process, the focus then shifts to determining the range of expert opinions for each criterion. This involves identifying the maximum and minimum values within the expert-weighted matrix, effectively capturing the spread of perspectives across the panel. The difference between these maximum and minimum values is then calculated, representing the level of disagreement or consensus among the experts for each criterion. This difference quantifies the degree of uncertainty or ambiguity associated with each criterion. These differences are

TABLE 8.1
Input data matrix

Expert	*M*	*F*	TSF	MRR	PR	*C*	Expert — years of experience	Expert-weights
E 1	8	6	6	7	4	5	4	0.0975
E 2	10	7	7	8	6	6	8	0.1951
E 3	9	10	5	9	7	5	3	0.0731
E 4	8	6	4	8	7	4	9	0.2195
E 5	10	9	4	8	6	3	17	0.4146

TABLE 8.2
Weighted normalized input data matrix

Expert	*M*	*F*	TSF	MRR	PR	*C*
E 1	0.0173	0.0154	0.0225	0.0171	0.0130	0.0212
E 2	0.0434	0.0359	0.0525	0.0390	0.0390	0.0509
E 3	0.0146	0.0193	0.0141	0.0165	0.0171	0.0159
E 4	0.0390	0.0347	0.0338	0.0439	0.0512	0.0382
E 5	0.0921	0.0982	0.0638	0.0829	0.0829	0.0541

TABLE 8.3
Criteria weightage as computed by the CIMAS method

Parameter	*M*	*F*	TSF	MRR	PR	*C*
$R_{j\,max}$	0.0921	0.0982	0.0638	0.0829	0.0829	0.0541
$R_{j\,min}$	0.0146	0.0154	0.0141	0.0165	0.0130	0.0159
B_j	0.0775	0.0828	0.0497	0.0665	0.0699	0.0382
L_j	0.2015	0.2153	0.1293	0.1728	0.1818	0.0993

subsequently used to compute the ranking value, denoted as Lj, which represents the final weight assigned to each criterion. This weight reflects both the inherent importance of the criterion and the level of agreement among experts regarding its significance. These calculations are documented in Table 8.3, following the formulas and equations previously outlined in the CIMAS model section.

The final stage of the CIMAS method focuses on validating the reliability of the calculated criteria weights. This validation involves a crucial feedback loop, where the experts are once again consulted. Each expert is asked to individually assign weights to each criterion, expressed as percentages ranging from 0% to 100%. These percentages must collectively total 100% across all criteria. This step serves as an independent check on the CIMAS-derived weights, assessing the level of consistency between the model's output and the direct assessments provided by the experts.

To measure this consistency, an RI is calculated using Equation 8.8. In this case, the RI value of 0.0708 falls significantly below the predefined acceptable threshold of 0.1. This outcome indicates a high level of agreement between the CIMAS-derived weights and the experts' direct evaluations, reinforcing the reliability and dependability of the entire weight determination process. This rigorous validation step ensures that the final weights used for NTM process selection are robust, consistent, and genuinely reflective of expert judgement.

8.4.2 Prioritization of NTM Processes through the PIV Approach

The next step in the NTM process selection for machining titanium involves utilizing the PIV method to conduct a systematic evaluation and ranking of potential NTM candidates. This phase begins with the identification of suitable NTM processes from a previously compiled list of eight recognized techniques. By starting with these options, the evaluation process ensures a thorough assessment of viable choices for machining titanium. Once the potential processes are identified, a detailed comparative analysis is carried out. This analysis involves characterizing each NTM process based on the six predefined criteria. Quantitative performance metrics, reflecting each process's capabilities and limitations with respect to each criterion, are systematically compiled. This comprehensive performance data, presented in Table 8.4, serves as the initial decision matrix.

Following the guidelines of the established PIV methodology, the next step involves converting the raw performance data into a normalized and weighted decision matrix. This transformation is crucial for enabling objective comparisons

TABLE 8.4
Initial decision matrix

NTM process	MRR	*F*	*M*	TSF	PR	*C*
AJM	0.8000	1.0000	4.0000	2.5000	0.2200	1.0000
USM	300.0000	1.0000	4.0000	1.0000	2.4000	2.0000
CHM	15.0000	1.0000	4.0000	3.0000	0.4000	3.0000
EBM	1.6000	4.0000	4.0000	2.5000	0.2000	4.0000
LBM	0.1000	4.0000	4.0000	2.0000	1.4000	3.0000
ECM	1500.0000	5.0000	4.0000	3.0000	100.0000	5.0000
EDM	800.0000	1.0000	5.0000	3.5000	2.7000	3.0000
PAM	75000.0000	5.0000	4.0000	5.0000	50.0000	1.0000

TABLE 8.5
Normalized decision matrix

NTM process	MRR	*F*	*M*	TSF	PR	*C*
AJM	0.0000	0.1078	0.3417	0.2931	0.0020	0.1162
USM	0.0040	0.1078	0.3417	0.1172	0.0215	0.2325
CHM	0.0002	0.1078	0.3417	0.3517	0.0036	0.3487
EBM	0.0000	0.4313	0.3417	0.2931	0.0018	0.4650
LBM	0.0000	0.4313	0.3417	0.2345	0.0125	0.3487
ECM	0.0200	0.5392	0.3417	0.3517	0.8939	0.5812
EDM	0.0107	0.1078	0.4272	0.4103	0.0241	0.3487
PAM	0.9997	0.5392	0.3417	0.5862	0.4469	0.1162

between NTM processes, taking into account potential differences in measurement scales and the relative importance of various criteria. Initially, the decision matrix, which contains the raw performance metrics for each NTM process across all criteria, undergoes a normalization procedure. As illustrated in Table 8.5, this process standardizes the data onto a comparable scale, helping to minimize any biases introduced by varying units or magnitudes of measurement. This approach ensures that all criteria are evaluated equitably, regardless of their original measurement scales.

After normalization, the decision matrix undergoes further refinement through a weighting process. This stage, as shown in Table 8.6, incorporates the previously established criteria weights, ensuring that the relative importance of each criterion is accurately reflected in the final evaluation. By integrating these weights, the weighted normalized decision matrix offers a comprehensive and objective representation of each NTM process's relative performance, considering both their inherent capabilities and the prioritized significance of each evaluation criterion. This refined matrix acts as a crucial input for the final ranking and selection of the most suitable NTM process for machining titanium.

The next step in the PIV method involves calculating the proximity of each NTM alternative to an ideal solution. This is done by determining a Proximity Index,

TABLE 8.6
Weighted normalized decision matrix

NTM process	MRR	*F*	*M*	TSF	PR	*C*
AJM	0.0000	0.0232	0.0689	0.0379	0.0004	0.0115
USM	0.0007	0.0232	0.0689	0.0152	0.0039	0.0231
CHM	0.0000	0.0232	0.0689	0.0455	0.0007	0.0346
EBM	0.0000	0.0929	0.0689	0.0379	0.0003	0.0462
LBM	0.0000	0.0929	0.0689	0.0303	0.0023	0.0346
ECM	0.0035	0.1161	0.0689	0.0455	0.1625	0.0577
EDM	0.0018	0.0232	0.0861	0.0530	0.0044	0.0346
PAM	0.1728	0.1161	0.0689	0.0758	0.0813	0.0115

TABLE 8.7
Proximity index for NTM alternative

NTM process	MRR	*F*	*M*	TSF	PR	*C*	Overall proximity value	Rank
AJM	0.1728	0.0929	0.0172	0.0227	0.0000	0.0000	0.3056	5
USM	0.1721	0.0929	0.0172	0.0000	0.0036	0.0115	0.2973	4
CHM	0.1727	0.0929	0.0172	0.0303	0.0003	0.0231	0.3365	7
EBM	0.1728	0.0232	0.0172	0.0227	0.0000	0.0346	0.2706	3
LBM	0.1728	0.0232	0.0172	0.0152	0.0020	0.0231	0.2534	2
ECM	0.1693	0.0000	0.0172	0.0303	0.1622	0.0462	0.4252	8
EDM	0.1709	0.0929	0.0000	0.0379	0.0041	0.0231	0.3288	6
PAM	0.0000	0.0000	0.0172	0.0606	0.0809	0.0000	0.1588	1

which serves as a quantitative measure of how closely each alternative meets the desired performance characteristics across all criteria.

This calculation, grounded in the established PIV methodology, involves a multi-step process. First, the weighted proximity indexes are computed, capturing the relative performance of each alternative with respect to each criterion, while accounting for the varying importance of those criteria. These weighted proximity indexes are then aggregated, as per Equation 8.13, to derive the overall Proximity Index (d_i) for each alternative (A_i). This index provides a comprehensive measure of the relative desirability of each NTM process, as shown in Table 8.7, facilitating a clear and objective ranking based on their proximity to the ideal solution. The final stage of the PIV method involves ranking the NTM alternatives based on their calculated Proximity Indexes (d_i). This ranking, performed in ascending order of d_i values, prioritizes alternatives with the smallest index. This approach effectively identifies the NTM process that exhibits the closest proximity to the ideal solution, as indicated by the lowest d_i value. Therefore, the alternative with the smallest d_i emerges as the most preferred choice for machining titanium, signifying its superior performance across the weighted criteria and its optimal alignment with the desired characteristics.

The combined CIMAS-PIV methodology used for selecting the most appropriate NTM process for titanium has demonstrated promising outcomes. The analysis highlighted PAM as the optimal NTM process, a conclusion that aligns with the findings of Chakladar et al. (2009). This consistency in results, achieved independently through different methodologies, reinforces the reliability and credibility of the integrated CIMAS-PIV approach.

The efficacy of the method was further validated through a comparative analysis with the rankings obtained by Chakladar et al. (2009), revealing a strong correlation. The calculated Spearman's rank correlation index of 0.78 demonstrates a significant degree of agreement between the two studies. This high correlation coefficient underscores the relevance and effectiveness of the integrated CIMAS-PIV method in practical scenarios, showcasing its capability to generate reliable and consistent results that align with established research in the field. This outcome emphasizes the potential of the integrated method as a robust decision-making tool for NTM process selection, offering a systematic and validated approach for optimizing machining processes.

8.5 CONCLUSION

This research effectively tackles the critical challenge of selecting the optimal NTM process for titanium, a material widely used in high-performance sectors such as aerospace, biomedical engineering, and military. By combining the CIMAS method with the PIV technique, the study introduces a robust hybrid framework that integrates expert judgement with objective analysis, ensuring practical applicability and methodological rigour.

The findings indicate that PAM is the most suitable NTM process for machining titanium, closely followed by LBM. PAM's advantages include a high MRR, capability to manage intricate geometries, and minimal thermal impact, making it highly effective. Additionally, the study highlights geometric precision, represented by the F criterion, as the most influential factor in process selection, whereas cost considerations hold less weight in high-value applications.

The integration of CIMAS and PIV methods demonstrates strong reliability and consistency, with a notable alignment to existing literature, reinforcing the framework's robustness and adaptability. Beyond its application to titanium, the methodological innovations presented in this study offer potential for guiding NTM process selection for other difficult-to-machine materials, expanding its relevance to a variety of manufacturing contexts.

On a broader level, this research makes a significant contribution to the field of MCDM by offering a structured and flexible approach that connects theoretical models to real-world industrial needs. Future research could further develop this framework by assessing its performance in dynamic and uncertain decision-making environments, leveraging real-time data and advanced computational tools to enhance its functionality. By addressing both technical and managerial aspects of process selection, this study lays a strong foundation for continued progress in manufacturing optimization.

REFERENCES

Biswas, S., & Anand, O. P. (2020). Logistics competitiveness index-based comparison of BRICS and G7 countries: An integrated PSI-PIV approach. *IUP Journal of Supply Chain Management*, *17*(2), 32–57.

Bošković, S., Jovčić, S., Simic, V., Švadlenka, L., Dobrodolac, M., & Bacanin, N. (2023). A new criteria importance assessment (Cimas) method in multi-criteria group decision-making: Criteria evaluation for supplier selection. *Facta Universitatis, Series: Mechanical Engineering*, 23(2), 335–349.

Bouraima, M. B., Jovčić, S., Švadlenka, L., Simic, V., Badi, I., & Maraka, N. D. (2024). An integrated multi-criteria approach to formulate and assess healthcare referral system strategies in developing countries. *Healthcare Analytics*, *5*, 100315.

Chakladar, N. D., Das, R., & Chakraborty, S. (2009). A digraph-based expert system for non-traditional machining processes selection. *The International Journal of Advanced Manufacturing Technology*, *43*, 226–237.

Chakraborty, S., Dandge, S. S., & Agarwal, S. (2020). Non-traditional machining processes selection and evaluation: A rough multi-attributive border approximation area comparison approach. *Computers & Industrial Engineering*, *139*, 106201.

Chatterjee, P., Mondal, S., Boral, S., Banerjee, A., & Chakraborty, S. (2017). A novel hybrid method for non-traditional machining process selection using factor relationship and Multi-Attributive Border Approximation Method. *Facta Universitatis, Series: Mechanical Engineering*, *15*(3), 439–456.

Goswami, S. S., Mohanty, S. K., & Behera, D. K. (2022). Selection of a green renewable energy source in India with the help of MEREC integrated PIV MCDM tool. *Materials Today: Proceedings*, *52*, 1153–1160.

Kara, K., Yalçın, G. C., Çetinkaya, A., Simic, V., & Pamucar, D. (2024a). A single-valued neutrosophic CIMAS-CRITIC-RBNAR decision support model for the financial performance analysis: A study of technology companies. *Socio-Economic Planning Sciences*, *92*, 101851.

Kara, K., Yalçın, G. C., Kaygısız, E. G., Simic, V., Örnek, A. Ş., & Pamucar, D. (2024b). A picture fuzzy CIMAS-ARTASI model for website performance analysis in human resource management. *Applied Soft Computing*, 162, 111826.

Kara, K., Yalçın, G. C., Simic, V., Korkuç, Ç., Çiçek, İ., Afacan, E., & Pamucar, D. (2024c). An interval-valued spherical fuzzy CIMAS-WISP group decision-analytic model for blockchain platform selection in digital projects. *Applied Soft Computing*, 111810.

Khan, N. Z., Ansari, T. S. A., Siddiquee, A. N., & Khan, Z. A. (2019). Selection of E-learning websites using a novel Proximity Indexed Value (PIV) MCDM method. *Journal of Computers in Education*, *6*, 241–256.

Kumari, A., & Acherjee, B. (2022). Selection of non-conventional machining process using CRITIC-CODAS method. *Materials Today: Proceedings*, *56*, 66–71.

Nguyen, H. S., Hieu, T. T., Thang, N. M., Tan, H. N., Can, N. T., Thao, P. T., & Bao, N. C. (2024). Selection of crankshaft manufacturing material by the PIV method. *Engineering, Technology & Applied Science Research*, *14*(4), 14848–14853.

Prasad, K., & Chakraborty, S. (2014). A decision-making model for non-traditional machining processes selection. *Decision Science Letters*, *3*(4), 467–478.

Roy, M. K., Das, P. P., Mahto, P. K., Singh, A. K., & Oraon, M. (2021). Non-traditional machining process selection: A holistic approach from a customer standpoint. In K. Kalita, R. K. Ghadai, & X. Gao (Eds.), *Data-Driven Optimization of Manufacturing Processes* (pp. 165–178). IGI Global.

Sadhu, A., & Chakraborty, S. (2011). Non-traditional machining processes selection using data envelopment analysis (DEA). *Expert Systems with Applications*, *38*(7), 8770–8781.

Saluja, R. S., Mathew, M., & Singh, V. (2023). Improved proximity indexed value MCDM method for solving the rank reversal problem: A simulation-based approach. *Arabian Journal for Science and Engineering*, *48*(9), 11679–11694.

Sampath, B., Sasikumar, C., & Myilsamy, S. (2023). Application of TOPSIS optimization technique in the micro-machining process. In M. A. Mellal (Ed.), *Trends, Paradigms, and Advances in Mechatronics Engineering* (pp. 162–187). IGI global.

Talib, F., & Asjad, M. (2019). Prioritisation and selection of non-traditional machining processes and their criteria using analytic hierarchy process approach. *International Journal of Process Management and Benchmarking*, *9*(4), 522–546.Trung, D. D. (2021). The combination of Taguchi–Entropy–WASPAS–PIV methods for multi-criteria decision making when external cylindrical grinding of 65G steel. *Journal of Machine Engineering*, *21*(4), 90–105.

Index